T0186446

A Perspective on
U.S. Farm Problems and
Agricultural Policy

About the Book and Authors

Today's American farmers struggle constantly with a
financial crisis. News reports chronicle the dreary events--
foreclosures, protests, governmental debates, distress sales.
There is great pressure on the U.S. government to ease the
current farm financial situation, but there is opposing pressure to wean the farm sector from government support, to
move agriculture toward a market-oriented economy, and above
all, to reduce government outlays. Because the financial
crisis permeates the agricultural community, critical choices
must be made regarding the role of government in the farm
sector, and alternate policies must be conceived.

A Perspective on U.S. Farm Problems and Agricultural
Policy provides a framework for evaluating national policy
alternatives and attempts to improve our understanding of
the nature of the farm sector and its problems. The discussion covers important issues such as farm size, the process
of dynamic adjustment to changing technology, and financial
disruptions associated with periodic land booms and busts.
The issues are presented in historical perspective and in
relation to past agricultural policies.

The authors move to more specific examination of the
current financial crisis by looking at various sizes and
types of midwest farms. They establish different categories
of financial stress in the context of both historical data
for the national farm sector and more recent data on individual farms, drawing conclusions about the primary determinants of financial stress and the manner in which individuals are adjusting to it.

Lance McKinzie is currently an analyst with the State
Utility Forecasting group in the Department of Agricultural
Economics at Purdue University. Timothy G. Baker is an
associate professor of agricultural economics at Purdue University. Wallace E. Tyner is a professor of agricultural
economics at the same institution.

A Perspective on
U.S. Farm Problems and
Agricultural Policy

Lance McKinzie, Timothy G. Baker, and Wallace E. Tyner

Routledge
Taylor & Francis Group

LONDON AND NEW YORK

First publishing 1987 by Westview Press, Inc.

Published 2018 by Routledge
52 Vanderbilt Avenue, New York, NY 10017
2 Park Square, Milton Park, Abingdon, Oxon OX14 4RN

Routledge is an imprint of the Taylor & Francis Group, an informa business

Copyright © 1987 Taylor & Francis

All rights reserved. No part of this book may be reprinted or reproduced or utilised in any form or by any electronic, mechanical, or other means, now known or hereafter invented, including photocopying and recording, or in any information storage or retrieval system, without permission in writing from the publishers.

Notice:
Product or corporate names may be trademarks or registered trademarks, and are used only for identification and explanation without intent to infringe.

Library of Congress Catalog Card Number: 86-051501
ISBN 13: 978-0-367-01387-5 (hbk)
ISBN 13: 978-0-367-16374-7 (pbk)

Contents

Tables

Figures

Preface

The nation now faces critical choices regarding the role which government can and should play in the farm sector. There is great pressure for our government to act to ease the current farm financial situation. There is also great pressure to wean the farm sector from government support, to move agriculture toward a market oriented economy, and above all to reduce government outlays. The purpose of this book is to provide a perspective for evaluation of these choices.

This book represents an attempt to improve our understanding of the nature of the farm sector, its problems and national policy alternatives. Important issues such as farm size, the process of dynamic adjustment to changing technology and financial disruptions associated with periodic land booms and busts are covered. These issues are presented in an historical perspective and in relation to past agricultural policies.

The book also provides an in-depth view of the current financial crisis situation on different sizes and types of midwest farms. Different categories of financial stress are explained in the context of historical data for the national farm sector and of more recent data on individual farms. Conclusions are drawn about the primary determinants of financial stress and the manner in which individuals are adjusting to this stress.

A fundamental problem with the farm sector is defined in the book as the condition responsible for the trauma which must be endured each time beliefs about the level of future returns to farm assets are changed and substantially lower returns are expected. The problem occurs in the aftermath of each land boom. The effect of this change in beliefs is a large transfer of wealth to future consumers of agricultural outputs and from current owners of farm assets. That transfer results in great trauma, first because it is quite concentrated on farmers and second because necessary adjustments are further concentrated among those farmers with substantial debt financing.

Traditional commodity programs do not make the farm sector more resilient in the face of changes in beliefs about farm returns nor do traditional credit programs. It is argued in this book that policy approaches capable of ameliorating this problem must encourage more equity financing and less debt financing or provide means whereby debt financing can be made more acceptable. The latter alternative is explored in detail.

A policy proposal is made with the objective of developing institutions which provide opportunity for individuals to transfer the risk associated with debt financing of farm assets to other individuals willing and able to accept that risk. The proposal provides both short- and long-term benefits to the sector:

- A corporate type institution would be developed to facilitate outside investment in farm assets.
- Current farm asset values would be supported as the above institution was initiated.
- A futures exchange for stock in the above institution would allow for transferring the risk associated with holding farm assets (especially those financed with debt).

Lance McKinzie
Timothy G. Baker
Wallace E. Tyner

Acknowledgments

Research behind the book and its writing was supported by the USDA through a cooperative agreement with
Purdue University. The idea for broadening traditional
patterns of food and agricultural policy research stemmed
from a conversation several years ago in Logan, Utah, with
Dr. Kenneth Clayton, who is now Director of the National
Economic Division of ERS. Ken felt strongly about the
need to view farm policy research in a time frame longer
than that fostered by the four-year farm bills. So much
research effort has been devoted in the past to preparing
for debate on the "next" farm bill that research seemed
all too confined to fussing with details of traditional
commodity programs. New thinking on old farm problems was
"too" infrequent. We were excited about searching for new
approaches. The cooperative agreement, several working
papers, and this book are the outcome.

Keith Schap, a freelance writer, performed something
of magic on our first draft. He went over the entire man-
uscript, changing a word here and rewriting a section
there, managing to greatly clarify our treatise and often
our own thinking without distorting our original meanings.
Keith is a fine writer. Tana Taylor, Jeri McIntyre, and
Dixie Strubel shared orchestration of the word processing.

1

Introduction

American farmers of the mid-1980s are struggling with an enormous financial crisis. Lenders and businessmen who primarily serve the farm community face similar distress. Hardly a week passes without news reports chronicling yet further dreary episodes in this unhappy saga--foreclosures, protests, governmental debates, distress sales.

Worse, no one seems to know what to do about all of this. Some propose continuing price supports for this or that critical commodity. But the problem is not peculiar to corn farmers or strawberry growers, cattlemen or cotton farmers. The financial crisis prevails throughout the agricultural community.

Some observers think farmers should take their chances in the marketplace like other businessmen. They entered farming to make a profit with a full awareness of the risks, those observers contend, and when adverse situations arise, they should either discontinue their venture or not--like people in other fields.

In some important ways, though, farmers differ from many businessmen. Any attempt to understand their dilemma or find a solution for it has to start with that assumption. For farmers face their risks alone. The typical farmer, facing the need for extra capital, does not issue more stock as a typical manufacturer does. He tends not to enter into limited partnerships as venturers in oil exploration and other kinds of real estate often do. His source of money tends to be his bank, or a similar lender. And all he has to bank on are his land holdings and his potential crops. All for good reason. Farmers are independent souls--often that is a virtue. Here it may be a flaw. More important, agricultural policy and social setting have made this so. People have opposed the idea of

1

corporate farms for some reason. And policy directions which create cheap interest and support commodity prices have urged this approach to farm finance. If all does not go well in terms of land values or yields, he faces his debts alone.

And debt is a major part of the problem in agriculture. The USDA, with its sure statistical grasp, defines the problem in terms of leverage problems. Basic is the notion that a businessman must finance all assets--either with equity or with debt. Thus, assets equal equity plus debt. For a farmer, the primary asset is land, and that is a relatively stable factor. A farmer is unlikely to be able to alter that holding quickly or substantially. To increase profits, the farmer must typically add debt, or use leverage. Leveraging multiplies profit, but it also multiplies risk.

If a farmer finances, say, $100,000 worth of assets with $30,000 in equity and $70,000 in debt, then his debt-asset ratio is 70%. If he wishes to increase profits, he might decide to assume more debt, perhaps $120,000. Now his debt-asset ratio has risen to 80%. When all goes well, that will dramatically boost the profits of his operation. If his return holds at five percent, his profit will rise from $5,000 to $7,500. Conversely, if crops are poor, prices go down, or land values plummet-- all conditions which have plagued farmers in recent years --then the farmer realizes how dramatically his risk has increased. And he may well have severe problems dealing with that risk.

In fact, according to the USDA, as many as 145,000 farms may have extreme financial leverage problems (the USDA defines "extreme" in terms of debt-asset ratios greater than 70%). As many as 243,000 farms may have serious financial leverage problems (debt-asset ratios between 40% and 70%). That is, almost 16% of the 2.5 million farms are experiencing serious or extreme leverage problems.

Following developments between 1980 and 1985 makes the problem seem even more severe. Consequences have not yet worked through the systems. Apparently farmers will continue to feel distress for some time to come. Rates of failure may well increase.

The number of farms experiencing extreme leverage problems rose during those years by 94 percent, and the number with serious leverage problems rose by 26 percent. The average net worth for very highly leveraged farms ranged from trivial to negative across all sales classes.

2

This indicates that about half of these farmers (72,000) were technically insolvent by January 1985. Cash short-falls range between 12 percent and 33 percent of net worth for farms with debt asset ratios between 40 and 70 per-cent. Cash shortfall and equity conditions suggest that few farms in the highly leveraged category could survive more than three or four additional years unless conditions change, and almost none of the very highly leveraged farms could continue more than two years. While less highly leveraged farmers fared better, as many as 85 percent of all farms experienced cash shortfalls in 1983.

Financial stress among farmers also directly concerns agricultural lenders. As of January 1985, ag lenders faced losses on approximately 12 percent of their loans. If the trend were to continue two or three more years past that point, the proportion of technically insolvent loans would double at least to 24 percent. Many lenders will have to face the possibility of failure and reorganization if more than a few percent more of their borrowers default on their loans. So the financial stress extends to the agricultural credit sector.

On the surface, the fundamental financial problem for many farmers appears to be related to capital investments. From the current vantage point, it seems clear that a great many farmers invested in assets that were overpriced and used too much debt financing to do so. The present high interest rates only compound the problem. However, it is unlikely that anyone--farmer, banker, extension advisor, farm consultant, or university economist--would have thought the investments bad ones on the day they were made. At least, few of these people were issuing words of caution. After all, during a thirty year period, our eco-nomic community issued strong signals to farmers to use financial leverage and to invest aggressively. Underlying those signals were a variety of assumptions: land and exports would continue to appreciate, commodity prices would continue to increase or at least hold steady, the financial climate--interest rates, inflation, and so on--would follow present trends.

Suddenly economic conditions changed. Society elected to end rapid inflation. The financial community adjusted money supplies and values accordingly. Exchange rates altered markedly. Interest rates responded. And the agricultural sector found that the assumptions on which it had made those investments were no longer valid. As a result, what had seemed a reasonable action was now a "bad investment." Suffering farmers heard from all sides

3

that they were simply paying the price for bad management. Clearly, the situation they found themselves in was far more complex than it seemed on the surface.

A more probing examination of the agricultural sector's financial problems reveals that they follow from a deeper historical source. The farm sector is readjusting to the disruptive effects of the boom period of the 1970s. Actually, agriculture is currently more profitable than it was during the 1950s and 1960s. Estimates of returns to agricultural assets show that they exceed those characteristic of the pre-land boom period. But the quantity of new debt which farmers assumed during the 1970s and the recent high real interest rates result in an increase in the cost of debt which offsets the increase in returns to assets.

In general, land booms bring about increased debt. Substantial land transfers take place at prices which changing economic conditions may later invalidate. That is, society invalidates the assumptions underlying those conditions, these now bad investments bode ill for the sector as a whole in spite of the fact that agriculture is currently relatively profitable. Returns to management and equity are on the whole about zero.

Numerous observers of the agricultural scene attribute the financial troubles of certain farmers to bad management. Although that may be true in some cases, in many others, the most severe difficulties appear to visit those who in other times would be thought to be among the shrewdest of managers. Financial conditions vary greatly among individuals, but there are some broad generalizations which obtain:

- Farmers with little or no debt are fairing quite well after the boom period with much new wealth and higher income returns to assets.
- Farmers with average debt, average return on assets and paying an average rate of interest have about zero return to management and equity.
- Farmers with higher than average proportions of debt, but earning an average return on assets and paying an average rate of interest, are suffering negative returns to management and equity.

Those who acquired high levels of debt during the boom period may well be some of the better entrepreneurs and managers, although it is possible to state after the fact that they should have known better than to assume land

4

values would continue to increase. It is possible to argue that these people were simply born at the wrong time.

Interestingly, this situation of extreme financial stress in the agricultural sector of our economy is not unprecedented. There have been several parallel series of events whose patterns are remarkably similar to what we are seeing now. Working backwards in time, there were land booms followed by severe busts during the Great Depression of the 1930s, in the 1870s, and following the Napoleonic Wars in the 1820s.

These events suggest a set of interesting conclusions. If a problem situation is a chance occurrence--a one-time circumstance, then some kind of intervention to tide people through the crisis period makes sense. In general, that is how governmental agencies and others concerned with farm policy are reacting at present. But, if the events seem to be part of a larger historical sequence, then such intervention will never really work for it treats only superficial aspects and fails to come to grips with the real dilemma. Many people, of all political and economic persuasions, implicitly acknowledge this when they admit that price supports or production controls do not really produce satisfactory results.

The problem, then, must be structural. Its roots must derive from factors inherent in the institution--in this case, the institution of agricultural finance. When that is true, then there is an illusory aspect to success. Lacking a sound institutional basis, apparent success is a fragile thing. At present, the remarkable rise and fall of a number of agricultural entrepreneurs lends support to those conclusions. Those people ran their farms just like the best farm business thinkers said they should. During the boom, theirs was a phenomenal success. When the situation changed, though, they were among the first to fall, and theirs was a colossal fall.

Nevertheless, recognition of a historical pattern in these events provides hope that we can extract a solution for the financial plight of farmers which will be more than a stop-gap. The historical pattern points up structural problems in our system of farm finance. Accordingly, we should look for an institutional correction rather than satisfying ourselves with treating symptoms as we have previously tried to do.

Given the severity of these problems, the historical precedent and the widespread dissatisfaction with traditional policy answers, we wonder why policy makers and agricultural financiers have not recognized this. Two reasons suggest themselves immediately.

First, we tend to focus on the present and tend to lack historical perspective. These boom-bust sequences are long, spanning two or three generations. And they are separated in time so that the actual crises are even farther apart. As a result, grandsons lack real awareness of what their grandfathers may well have learned from experience. Each crisis becomes an occasion to reinvent a financial wheel.

Second, and probably more basic, we seem to have discovered a situation where people are trying to operate using an old, worn out conceptual framework--or paradigm. And to attain any real solution will require a reorientation of our thinking--adopting a new paradigm.

Policy makers, financial people, farmers--everyone, in fact, with a share of concern in the matter of farm finance--tend to see the farm financial problem in terms of income problems. Given that income is his problem, then any of the traditional policy responses should provide an adequate solution. But they do not. Commodity price supports, for instance, may help farmers cope with an immediate crisis, but they encourage more debt, they invite expansion where most other market signals urge otherwise, they create trading problems in both domestic and overseas markets, and they undermine the stability of banks and other businesses that serve the farm sector. In short, in exchange for a minuscule temporary benefit these policies create numerous long lasting problems. So it is with virtually every policy approach.

Actually, that describes well what happens when people in any science try to solve problems in the context of a worn out paradigm. The attempt only leads to a bewildering array of new problems. As we argue that the real financial problem for farmers concerns asset values, not income, we are in effect urging adoption of a new paradigm. For we think that once all of us with concerns in this area reconceptualize the farm finance problem along the lines we suggest, an interesting approach to a solution will suggest itself--one which promises farmers both short and long term aid and seems to avoid the unpleasant side effects of traditional policies.

This, then, is the thrust of our discussion--to reconceptualize the farm financial problem by concentrating on the basic issue of asset values rather than on the secondary matter of income and to follow the logic of that

6

realignment of our thinking to develop a policy proposal which provides farmers with an institutional mechanism which holds out the promise of actually coping with the root financial problems.

A Brief Perspective on Adjustment Problems

Finding an appropriate solution for the present crisis is a difficult task--conceptually and practically. In part, the solution we derive will follow from our understanding of the problem. Since a significant aspect of the problem involves the farm sector's approach to adjusting to change, it is well to review briefly the characteristic patterns and assumptions which shape those adjustments.

In our view, farm sector adjustment entails: 1) ongoing growth and adjustment to technological development, and 2) immediate adjustment to major disruptions in price expectations. Each aspect keys on certain central issues and assumptions. The farm economy has responded to pressure to grow and adapt to new technology by substituting capital for labor and consolidating smaller farms into larger units--all in search of higher returns for management and entrepreneurial skill. The issues surrounding responses to major disruptions in price expectations include substantial transfers of wealth among individuals and misguided capital investments. The land transfers causing the most severe financial stress, ironically, involve typically the young, aggressively expanding farmers and those who seem to be progressive individuals with above average management skills.

Overall, this view of the adjustment patterns allows us to accommodate a variety of causes and effects. The resulting conceptual richness provides for the incorporation of diverse events into a general understanding which has solid intuitive appeal and which creates a basis for an interesting framework for policy formation.

Each aspect of this has different roots and different effects on the structure and well being of the farm economy. Viewed in this manner, observations of different events in the sector, including the current situation, fit into a general understanding which has solid intuitive appeal and represents the beginning of a framework which will facilitate policy formulation. Brief introductory comments on both of these parts follow.

Ongoing Adjustment

People, affiliated with agriculture or not, often focus on the highly technical aspects of farming--crop production, machinery selection, operation, and maintenance. But farming is a business. Therefore, the equally technical factors of business operation, planning, and finance are profoundly important. We take the approach (suggested by Madden, 1984) that systematic variations in returns which are so strongly associated with size follow largely from management and entrepreneurial ability. Given that view it seems eminently reasonable that at any point in time, larger operations would tend to be operated by individuals possessing above average skills due to the selective mechanics of survival. We also view the farm firm as providing services, under the class of off-farm endeavor, in addition to agricultural product. This provides an explanation for the many smaller operations without resort to concepts of market imperfections and inefficiencies.

As the adjustment process unfolds, people substitute capital for labor, individuals leave the sector, and farms are consolidated under the control of fewer operators. Management skills seem to be available; the fact that in most years, farmers have competed strongly for land with which to expand their operations suggests an availability of management skills. Accumulated equity, in the hands of those with requisite management skills, has been an important constraint limiting more rapid expansion. Farm income is quite variable from year to year. For an individual farmer with a given preference for bearing risk, maintaining a desired probability of survival means limiting expansion to that directed by his accumulated equity. Our notion is that under "normal conditions" the market does a reasonable job of allocating residual returns between land, management skills, and other factors so as to regulate the structure of the sector under the ongoing process. We use the term "normal conditions" to mean that the general level of prices is relatively constant or changing in a manner consistent with expectations. But, normal conditions do not always hold, and this leads to the second issue.

Immediate Adjustments Associated with
Major Disturbances in Price Expectations

The second aspect of adjustment which we highlight concerns the disruptive effects of substantial and unanticipated changes in returns to farm assets. Though often correlated with changes in current returns, changes in beliefs about future returns necessitate the largest adjustments. The problem is that the structural organization of the farm sector, particularly our system of ownership and financing, does not readily accomodate downward changes in beliefs of the magnitude which occur from time to time. Large downward changes in beliefs about future returns cause great losses of wealth between farmers and signal many "bad" investments. Adjustments are necessary and must be made more quickly as the farmer's debt is higher. Many farmers must reduce the scale of their farm operations because they can no longer meet previously scheduled debt payments. Some farmers are forced to seek employment off the farm. Some stop farming altogether. A number become bankrupt. Even those not so severely affected must alter family consumption patterns--a widespread condition in the 1980s.

The structural condition responsible for so much trauma each time beliefs about future returns must be revised downward is a problem. The structure of agriculture, including government policies, encourages the use of debt financing. Much of the trauma is related to the concentration of debt within the sector. Under our system, those who are young or expanding the size of their farm operations carry the greatest share of debt, and the greatest share of the present financial burden.

Alternatives for the Future

Whenever a market is deemed to perform unsatisfactorily, policymakers tend to choose from two courses of action: (1) alter market outcomes or (2) treat the consequences. A third choice--one we find preferable--seldom receives consideration. The preferred choice is to develop new institutions or organizational arrangements which allow existing market forces to perform more effectively in serving society's desires. That action is preferable because there will, in general, be fewer undesirable side-effects with that approach. But, design of effective institutions is more difficult than direct interference. Both the problem and the behavior of the farm economy need to be well understood. While generally the more efficient means to a desired end, institutional innovation may not be a feasible approach in every situation. Nevertheless, we should consider this alternative carefully.

We think that it is possible to improve the system of ownership and finance of the farm sector to provide additional outside equity financing and to allow for the transfer of risks associated with changing farm asset values. Briefly, our proposal is to form a kind of corporate entity which would sell stock to the general public and then purchase and hold U.S. farm land as a profitable investment. The first effect of the entity would be to provide the farm sector with additional equity financing. It would also provide short run liquidity in the real estate market and perhaps forestall the potential failure of many of the farmers who have large mortgage debt.

In addition, the new institution would develop a market for trading in futures on the land holding entity's stock. The effect of this futures market would be to allow anyone, particularly those with large mortgage debts, to hedge against the risk of substantial declines in real estate values much as farmers now hedge crop prices. Such a hedge would be possible because the stock price would directly reflect land values. Broad declines in real estate values occur when people expect future returns to reach a much lower average level than previous beliefs have accounted for. Thus, participants in this market would hedge the long-run level of returns at the same time. Hedges of the long-run level of returns are not possible with currently existing commodity exchanges because contracts are short-term in nature. Also, mortgage debt on land covers returns over much longer horizons than that reflected in current commodity futures. On the other hand, one year futures on land values would reflect the returns from land over a period sufficiently long for the mortgage to be paid.

Organization of the Book

The following chapter presents a perspective on the farm economy which helps us to understand the different aspects of farm sector performance which have been addressed as farm problems in the past. The concept of size economies is particularly important in understanding the on-going process of growth and adjustment to technological progress. Size economies are closely linked with management abilities of individual farmers and, together, these issues contribute to explaining observed behavior toward the use of debt financing in the sector. Finally, these concepts are related to farm structure and the problem of farm financial crisis situations which this book addresses.

Chapters Three and Four provide information on the current financial situation in the farm sector and for different types and sizes of farms in the Midwest. Financial data on the farm sector are presented in a historical context to provide insight about the origin of the current financial crisis situation. Estimates are derived for the amount of equity which is drained from the farm sector by retiring farmers and other individuals leaving the industry. Summary data from two farm record keeping groups provide detailed information by type and size of farms and by level of debt. While debt is an important factor contributing to financial stress, these data illustrate that there is a great deal of variability in financial stress across farms which is not explained by the level of debt. Management or some other difficult to measure characteristic is extremely important.

The condition of the structure of agriculture which leads to great trauma from time to time is defined as a problem in Chapter Five. The trauma is associated with the bust part of land booms. Boom-bust periods and associated financial maladjustments are discussed in a historical context. We explore who gains and who loses during such a period. The nature of farm real estate markets is discussed in detail and related to the defined problem. Finally, the role of policy in addressing the problem is discussed.

Policy options and their likely consequences are discussed in Chapter Six. The problem is recast in terms of eleven axioms which provide a framework for discussion or argument. The axioms contain elements designed to provoke argument. We don't prove them. But they have solid intuitive feel and lead through dispensing with traditional programs toward adopting some policy measure designed to increase the amount of equity financing in the farm sector.

Discussion of policy options and their likely consequences highlights the fact that traditional programs do not address the basic problem defined in Chapter Five and, on the contrary, often worsen the situation. Of the options considered, only that which we propose addresses the immediate financial crisis situation, ameliorates the basic long-run problem and has no major undesired side-effects on other aspects of market efficiency and social welfare.

Chapter Seven combines a detailed discussion of the proposed option--a profit oriented land holding entity with organized trading in futures on the entity's stock. The chapter begins with a rationalization of the futures market for farm assets as a solution to the problem of farm financial crisis situations. Next the alternative mechanism for implementing such a futures market is presented--i.e., trading in futures on stock of a land holding entity. Implications of this proposal for the structure of agriculture are considered and, finally, limitations of the approach are listed.

2

A Perspective on Farm Structure

While the financial problems of the farm sector are severe, and cause intense suffering for many people, they need not occasion despair for our economy in general. The inverse of the notion <u>problem</u> is the notion <u>solution</u>. We take it as axiomatic that, serious as matters are in the farm sector, society can locate effective remedies.

However, the financial problem is a complex one which is superimposed upon phenomena of various kinds which people refer to collectively as the farm problem. Also, the entire matter involves a dynamic situation. Farmers adapt constantly to changing aspects of both farming and non-farm affairs. Accordingly, a quest for an understanding of the financial problems of the farmer must follow from a perspective on the structural organization of the sector and its adjustment dynamics.

Our perspective depends on a variety of assumptions concerning the concept of size economies, the role of management ability and entrepreneurial flair in farm business, and historical notions about fairness and social goals which have influenced policy over time.

Moreover, any definition of the structure of this economy must balance events on the sector level against those on the firm level. Sector wide, growth and technological adaptation are the major adjustment factors. Each firm adjusts to the stages in the career of the individual farmer. As we see how particular and general details interact, we can gain insight into the sector's needs for and use of debt financing.

Once we relate the concepts of farm size economy, the role of management ability, and the nature of the structure of this economic sector, we can gain the kind of

13

insight that motivates useful proposals concerning solutions.

Economies of Size and the Role of Management

In recent years, farm size has been a focus of attention. Often, this notion figures in debates concerning the relative virtue of the "corporate farm" as opposed to the "family farm". Without attempting to unravel the complexities, misunderstandings, and errors which obscure that discussion, we can observe that the technical concept of size economies in agriculture explains much about the farm sector, including the dynamic behavior. Various studies find cost to be inversely related to farm size until the level of "mid-sized" farms is reached. From that level, cost seems to be fairly constant, at least in the observed ranges (until management ability becomes limiting). That is, it costs a farmer with a small farm more to produce a given crop unit than it does his larger neighbors. But it is also true that, among the larger neighbors, when one attempts more than his management abilities allow, the cost begins to rise again. Income appears to increase directly with size throughout observed ranges; however we think management and entrepreneurial abilities serve to limit the range of farm size over which costs are constant. These, of course, differ for each individual.

Hawthorn (1919) reported farm size and operator management skills as key elements for attaining profitability when the same farms were surveyed repeatedly. Suter arrived at the same conclusions from his 1983 analysis of Purdue's Indiana Farm Accounts. Curiously, while both authors assigned primary significance to management, that aspect has received relatively little attention in the policy literature. However, many through the years have alluded to the importance of farm size in explaining farm income (see, for example, Warren (1928), Tweeten (1970), and Miller, et al. (1981)).

On Farm Size

We should not let recent heated discussion of farm size obscure our sense of the long history of this notion in our nation's economic development. In fact, farm policy and practical exigency have forced tension on this matter almost from the first days of the nation.

Technically, the relation between size and cost is straightforward. A simple unit cost curve can show the profitability of different sized farms. Tweeten (1970) and Miller, et al. (1981) develop unit cost curves which decrease up to a point as dollar volume of output increased and then tend to flatten out. Figure 2.1, which reveals Tweeten's estimates for 1960, illustrates such a curve.

"Unit" cost is the cost incurred per dollar's worth of output, and is measured on the vertical axis in these figures. A unit cost in excess of $1 implies or indicates a loss. However, cost in Figure 2.1 includes representative opportunity costs for capital and labor. Hence, in Figure 2.1 unit costs in excess of $1 imply that returns to these resources would be less than their opportunity costs. That is, returns to capital and labor are "low" in Figure 2.1 for farms with sales less than about $25,000. Depending on other factors, these farms may or may not actually experience loss.

The unit cost curve is neither an average cost curve nor a marginal cost curve. Rather it lies somewhere between those concepts. The curve represents "average cost" within an economic class of farm, but is a "marginal" cost curve among classes of farms (Tweeten, 1970, p. 178). The dotted line represents the marginal and average revenue (per dollar output) facing an individual farm. To the extent that included charges for capital and labor are representative of marginal opportunities for these resources, the positive or negative residual revenues represent returns to management or entrepreneurial efforts.

Most agricultural output is produced by a relatively few large farms whose activity significantly influences long-run price levels. However, most farms are smaller and, by conventional wisdom, bear unit costs somewhat above those of the larger farms. Tweeten (1970) understood the concept of decreasing costs and increasing returns to size as a general source of farm problems. He identified decreasing costs resulting from increasing farm size as the primary cause of low industry returns. While debate continues about the conceptually correct interpretation of these curves (see, for example, Center for Agricultural and Rural Development, 1984), the strong statistical association between income and farm size makes this

15

policy and practical relevance have little bearing on this
matter aside from the first days of the nation.

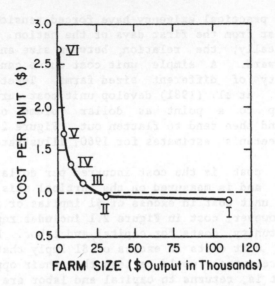

Figure 2.1. Long-run Unit Cost of Farm Production, by
Economic Class of Farms in 1960.

Reprinted from **Foundations of Farm Policy** by Luther
Tweeten by permission of University of Nebraska Press.
Copyright 1970 by the University of Nebraska Press.

important farm policy issue. Many small farmers evidently
earn low or negative returns to management and entrepreneurial ability while some large farmers earn positive
returns. Given Tweeten's suggestion that economies of
size are substantial, the large number of small farms with
low or negative returns to management causes income averaged over all farms to be low.

Of course, some small farms make efficient use of
resources. The lay of the land may dictate cultivation of
a number of small fields. Large equipment is definitely
not suitable on all of our agricultural land. As a
result, the depicted cost structure does not apply to all
situations. Further, such a situation is ideal for small
farms whose owners find primary employment in off-farm
endeavor.

But, consider resource configurations for which
larger operations would be feasible without higher unit
costs. Miller, et al., 1981, provide more recent estimates of economies of size relationships using 1978 USDA
cost of production data and a linear programming model.
Two different long-run average cost relationships are
illustrated in Figure 2.2. No charge for operator labor

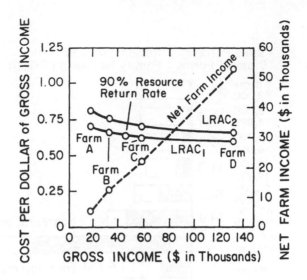

Figure 2.2. Relationship Between Constant or Decreasing Costs and Operator Net Income.

or management and no land rent or return is included in $LRAC_1$. A charge for the opportunity cost of operator labor is included in $LRAC_2$.[1] Miller, et al., estimate that most of the economies of size accrue to farms with just over $40,000 gross income and marked that size in Figure 2.2 as the point where fixed resources (mainly land in their view) earn 90 percent of the maximum potential rate of return. Given this kind of unit cost structure, operators' net farm income (which combines returns to labor, capital, risk, and management) tends to increase in a linear fashion as dollar volume of output increases.

The relationship between farm size and net farm income shows up dramatically in national farm income data for both good and bad years (see Figures 2.3 and 2.4). The group of relatively large farms has an average income several times the overall national average income. Thus, while they may have other problems, large farms cannot be viewed as having low family income. In addition, many very small farms do not have a family income problem due to non-farm related earnings. For all farms, total income (farm income plus off-farm income) averaged higher than the national average family income in both 1978 and 1981. Tosterud summed up the implications of these facts: "Six out of ten farms don't want, need, or deserve welfare" (1983, p. 932-4). Tosterud referred to the 900,000 or so

17

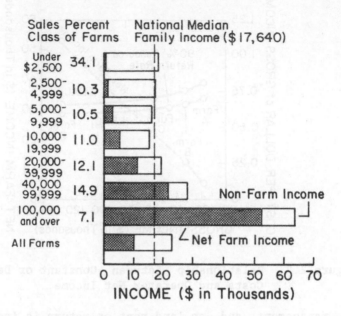

Figure 2.3.   Income Per Farm Operator Family, by Farm
              Size, 1978.

mid-sized farmers with below average incomes as being in a
transition zone.   For various reasons they are unable or
unwilling to generate sufficient off-farm income or to
grow large enough. Thus consideration of the relationship
among cost, size, and income helps us define these as
representing the core of the farm income problem.

     These analyses suggest a simple way of alleviating
the problems relating to a low average income in the farm
sector--simple on paper at least. Increasing the bimodal
farm size distribution by reducing the number of mid-size
farms would immediately raise average farm income. Mid-
size farmers with low income could either scale down their
farm operations and seek off-farm income or scale up operations to a satisfactory income producing level (McKenzie,
1978). Farm policy implications are not that straightforward, of course. Ours is not a totalitarian state. In a
given location, employment opportunities may not be available. And, most important, society may wish to maintain a
large number of small and medium sized farms in spite of
the fact that fewer farmers could produce the same amount.
Furthermore, the farm size issue is not static. A continual state of change makes policy prescription difficult.

18

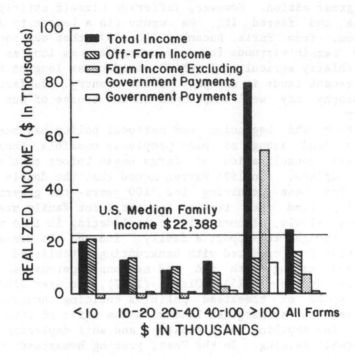

Figure 2.4. Farm Family Income, by Source of Income and Sales Class, 1981.

The Social Value of Numerous Small Farms

Thomas Jefferson's ideas about democracy provide the foundation for much that we value in our system. The Bill of Rights of the Constitution is only the most obvious example of his influence. Crucial for this discussion, Jefferson's theories stressed eradication of every form of aristocracy--abolition of every economic basis for landed estates, large holdings, class distinctions, and family prestige. That conception of democracy has served our country well (in the views of a broad spectrum of observers) and has become a deep-set tradition. Buttel and Flinn found that more than 80 percent of survey respondents agreed to the statement that the "family farm is very important to democracy" (1975). Jefferson believed that agriculture was a way of life, that tillers of the soil developed unique social virtues basic in the life of any great nation. However, Jefferson himself anticipated change and feared it. He wrote (in a letter to James Madison, from Paris, December 20, 1787) that our society would remain virtuous for many centuries, as long as they were chiefly agricultural, which would be as long as there were vacant lands in any part of the country. Jefferson's philosophy may well shape much of the course of our farm policy.

From the beginning our national policy has been to divide land among as many people as possible, though a contrary consolidation of farms began before settlement was complete. In 1928 Warren noted that the double process had been occurring for 100 years--the government dividing land into tracts too small for family use and farmers slowly, laboriously consolidating it into units large enough to support a family. Indeed, the Homestead Acts have been credited with bankrupting established farmers and making both old and new unprosperous (see for example Nourse (1925), Olsen (1927), and Grey (1938)). Grey spoke of homestead policies enticing thousands of families to submarginal land with the offer of farms too small for anything but a bankrupt and soil depleting kind of arable farming. In the West, grazing homesteads were, according to Grey, only one-fourth as large as the minimum required for a decent livelihood. Traditionally, then, two eminently reasonable and worthy goals seem to be in direct conflict as our farm economy interacts with our broader social goals: widespread control of resources and economically viable farms. Widespread control of resources is fundamental to our social system--a deep-set, basic value. "Once large blocks of land are consolidated under the control of one individual or family, history shows that it is difficult to break up such concentrations short of war or revolution" (Stanton, 1978, p. 735). And yet, given current technology, a viable sized (economically efficient) full-time family farm requires control of resources (e.g., dollar value of assets) far out of proportion to a family's share of the nation's total. No problem is simply economic. And no proposal which fails to recognize these interactions and tradeoffs can hope to provide a lasting solution.

A further idea, deeply rooted in traditional American conceptions of equality and law, is that farm wealth should be distributed equally among all children of a family. Thus, equity to finance a viable farm, in terms of size, must be divided once each generation under our present system. This further contributes to the existence

of small farm units with high leverage and has implications for each of the structural dimensions of agriculture
which are discussed in the section of this chapter on
Dynamic Adjustments. History shows that in many cases,
unless family members can arrive at a means to cooperate,
they will find it prudent to sell their land to other,
larger farmers (and so be out of farming altogether).
These tradeoffs are entwined in the current farm problems
and must be explicitly recognized.

On Management

Benedict tells of a western farmer's reaction to a
young government anthropologist:

> The farmer asked, 'Are you connected with the Depart-
> ment of Agriculture at Washington?' 'Yes, in a way,'
> replied the investigator. Then said the farmer,
> 'Young man, I want you to take this down in your
> notebook just as I'm saying it. You tell them jack-
> asses down there that farming is a business. It
> ain't a way of life' (1942, p. 476).

Those virtues peculiar to tillers of the soil, which
Jefferson extolled, are perhaps strongly related to the
fact that farming is a business. A farmer reaps what he
sows. He can work when and as he pleases, make his own
decisions, and accept all of the gains and losses from
that endeavor. While there is tremendous year to year and
decision to decision variability because of chance or
luck, some farmers excel over the long haul. Apparently,
some work harder, make better decisions, adopt profitable
new technology earlier, sell their output at higher
prices, and earn more income. Some farmers save more of
their income than others, and some of those use that
equity with various degrees of caution and prudence to
expand, applying their management skills to more
resources. Some farmers work less hard and perform less
successfully. Those earn lower incomes. The freedom,
responsibility and direct accountability must have in some
sense inspired Jefferson's faith in the strengths of
farmers.
At least since Jefferson, farming has been a highly
technical business. Jefferson himself experimented with
cultivation and conservation practices as well as new
machines. In the 1980s, the technical complexity of

farming and rate of change in that complex technology, is astounding. Farmers who spend too much time dreaming of a simpler life and make too many unscientific decisions will have lower incomes.

In his attempts to define successful farming, Tweeten (1970) found that certain choices concerning technical inputs produce significantly lower returns than others. Poor managers are those who use certain inputs to excess and others in too small quantities. Better managers seemed to allocate a greater proportion of expenditures for fertilizer, seed, high protein feeds, and pesticides, and lower proportion of expenditure for labor, machinery, and real estate. Further, a good manager will change the input pattern over time as new knowledge develops. Summary data from the 1983 Purdue Farm Records show that high profit farms used less fertilizer and machinery per acre and more pesticides than low profit farms, while obtaining higher yields and output prices. Some farmers have trouble striking the most useful balances among those inputs.

In any case, it is extremely difficult to quantify the impact of management skill on operator returns. Different farmers cannot operate the same land, at the same time, or under the same conditions in order to set up a good statistical experiment. Controlling for the many important factors is extremely difficult. However, it would be useful for policy-makers to have an estimate of the proportion of income variance across farms which is due to differing operator management skills in order to be able to design the best policy.

Management is an important factor to consider when interpreting the unit cost curves presented earlier. These curves fail to include a cost for management skills and entrepreneurial ability among operators. Taylor (1905) stated, "While there is no one proper size for farms in general, there is always a proper size of farm for a given man, at a given stage of his own development, on a given type of soil in a given line of production with given labor and market conditions" (p. 155). If there were an explicit market for a particular set of management skills, we would know their value. Or, if we could locate a cost for individual management skills in some other way, we could impute their cost. However, neither is likely to be possible in the near future.

Regardless of our ability to measure the effects of management skills in terms of unit cost curves, we know that for every farmer there is a level of output sufficiently large to strain his abilities to the extent that costs will rise because of inefficient management. At some higher level of output those increased costs will overtake the additional returns from expansion and any further incentive to grow will wither away. Because the unit cost curves offered so far assume management equally able to cope at various levels of size, one of the primary aspects of data which would cause the curves to turn up is absent. Chan, Heady, and Sonka (1976) have, though, estimated a long-run cost schedule for cash grain operations in which unit costs increase after about 800 crop acres (one to two man-years of labor). In general, it seems reasonable to expect that when farm size exceeds the farmer's ability to manage, costs will begin to rise. As Raup says:

> As farm size increases, management becomes a critical cost item. Management skills must be learned and producing a superior manager is expensive. To discuss the efficiency of farms without allowing for the differential costs of producing a manager, plus the costs of management error, feedback, and growth in skill is to ignore one of the most important aspects of transition in size of farm (1976, p. 1273).

Stanton (1978) noted that identifying what makes the curve fall had been easier than discovering evidence of a rising long-run cost after some point.

Heady considered limits to growth:

> The optimum size will differ between farms depending upon the stock of labor and management possessed in the household and it need not result in the most efficient use of a nation's resources (1967, p. 380).
> Continuance of the so-called family farm as the main structure of agriculture suggests, on the one hand, that if size economies exist, they soon give way to diseconomies. Concurrently, the continuance of small farms suggests the hypothesis that the economics of risk and uncertainty may be the final determinant of farm size in agriculture (1967, p. 350).

Heady and Krenz concluded "that farm businesses might not survive if they expanded acreage to a point equating marginal cost and revenue in an average year" (1962, p. 462). Stanton generalized and broadened this view:

23

It is not returns to capital, or returns to land, or
returns to any single scarce resource, that motivates
farm decisions ... Rather, it is some larger combination of things, including survival, net income over
time, enlarging the bundle of resources that the family controls, and increased prestige within the local
social system (1978, p. 735).

Dynamic Adjustment

Dynamic adjustment is on-going at two levels simultaneously. First, economic growth and technological change
are providing incentives for adjustment which effect the
entire sector. Second, each individual farmer is undergoing adjustment associated with his own life-cycle.

Economic Growth and Technological Change

Response to economic and technological incentives has
been continual for the sector as a whole with the idea of
an "appropriate" farm size for an individual a rapidly
moving target. Heady suggests that economic growth has
continually increased the supply of capital relative to
labor, making capital continually less expensive and causing a continual disequilibrium in agriculture (1967, Chapter 1). Perhaps so. From the beginning of our nation's
history farms have undergone rapid consolidation in search
of greater family income. In addition to the demise of
vacant land so feared by Jefferson, technological change
has worked steadily to reduce the number of farmers.
Labor-saving technological change has caused many people
to leave agriculture for higher incomes in off-farm
employment. Not only can fewer people successfully work a
given farm, the forces of changing technology have allowed
farmers to expand, ever increasing the economic size of a
family farm. In terms of the unit cost relationship,
technological gains tend to shift it to the right over
time and to extend the area over which the curve is flat.
That is, advancing technology allows a farmer to develop a
larger operation without increasing unit cost and so
develop more income. As a result of non-farm income
growth and labor-saving technology, Tweeten calculated
that farms needed to grow in size (value of output) at a

24

real rate of 5 percent during the three decades preceeding 1980 in order to maintain real operator returns to labor and management at constant levels (1980).

Madden, in an exhaustive review (1967), found that most studies indicated that farmers could achieve all economies of size with modern, fully mechanized one- or two-man farms. Though dated, the concept is probably still valid. But the measure is not very definitive. The rapidly increasing availability of larger machinery is illustrative. Stressing the concept, as opposed to the accuracy of the example, two men utilizing 6-row equipment can operate 750 acres of Indiana corn/soybean land with a given level of timeliness. Maintaining roughly the same timeliness and cost per acre, those same two men can operate 1,000 acres utilizing 8-row equipment, 1,500 acres utilizing 12-row equipment, and 2,000 acres utilizing 16-row equipment. Twenty-four-row equipment is, very recently, commercially available that would allow the two men plus some hired labor during harvest to operate over 3,000 acres. In each case, the machinery ownership and operation costs are roughly constant on a per acre basis. This provides support for the idea of unit cost curves which are flat over broad ranges of larger-sized farms. If it is valid to assume that the two men who are able to manage, successfully, a 750 acre farm are able to manage, successfully, a farm four times as large, this further illustrates how the magnitude of operator return increases directly with size.

The dynamic situation is complex. In the short-run there exist farms of differing size, and there is no apparent ideal size--no incentive to get larger or smaller. Any individual farmer who has the requisite managerial skill and is operating along the flat portion of his unit cost curve need only expand the operation's size to augment family income. And expand he will, but laboriously, subject to tremendous uncertainty about long-run price levels, exacerbated by the ever-present possibility of dramatically falling land values (the bust part of a land boom), limited by his ability to accumulate equity from retained earnings, and by his willingness to use financial leverage. The constant advance of technological progress--shifting the unit cost curve rightward and making the economic size of a family farm larger--and larger, perpetuates the growth in farm size.

25

Implications for Observed Levels of Farm Income

The dynamics of this phenomenon are an important factor underlying perceived farm problems. Since more and
larger equipment has become available each year, the flat
portion of the unit cost curve appears to be extending
rapidly to the right, moving over time toward higher levels of output. Increasing operator returns with farm size
causes tremendous pressure for the expansion of family
farms. Since larger farms produce a greater share of
total output than smaller ones. They will have substantial influence in determining long-run price levels.
Resource returns, particularly land rent, are continually
bid up as farm operators with high management ability view
additional expansions as profitable. Of course, when many
participants are attempting to expand and resources in
total are limited, competition eventually drives up the
prices of those resources for everyone. This raises the
costs for all participants, stressing particularly those
who, for whatever reasons, are unwilling to expand.

That raises a question concerning why do individuals
with farms so small as to earn unacceptably small incomes,
e.g., those with $10-$20,000 of gross sales in Figures 2.3
and 2.4, stay in farming? Don Paarlberg suggests (and we
agree) that behavior of farm people as entrepreneurs can
be better explained through taking account of opportunity
costs and psychic income. These vary enormously from person to person and are, perhaps, often more decisive than
our postulates from an accounting perspective. Many individuals undoubtedly prefer farming and don't want to do
anything else whatever the pecuniary income; but they do
not have the management ability necessary to successfully
operate larger enterprises. Some others may have no better opportunities. Tweeten (1970) referred to the impli-
cation of decreasing costs as being permanent low returns
for those farms operating in the decreasing portion of the
unit cost curve and permanent high returns accruing to
those whose size is above the level where all costs are
met. But, low returns would not necessarily need to be
permanent for specific individuals. It is possible that
many of the mid-sized farmers with below average incomes
are in transition: some intending to expand as conditions
permit and some intending to cut back and seek additional
off-farm endeavor. Future incomes will be higher for
these people. There may be a permanent, but open ended,
albeit transitory, class of mid-sized farms with low

family incomes due to structural conditions in the farm sector.

Implications for Supply Response

The phenomenon of high returns accruing to those whose size is above the level where all costs are met has important implications for the long observed tendency for aggregate supply not to respond to falling output price levels. In the short-run, falling price levels will move some farmers from the group earning high returns into the group earning low returns. But in the short-term, entre-preneurial ability can be a fixed resource and those firms earning positive returns to entrepreneurship have no incentive to leave the industry as output prices fall. They simply earn smaller returns. Output decreases only to the extent that some farmers earning low returns leave the industry and others, who are seeking higher returns to management and entrepreneurship through expansion do not replace them.

Implications for Farm Problems

Our major concern in regard to the on-going dynamic adjustments which are related to technological change and economic growth is with the increased capital requirements for the sector as a whole and with the even greater increases in capital requirement on a per capita basis. Substitution of capital for labor implies that more capital is utilized in the sector. But, in agriculture, labor also tends to be the owner of capital. When labor leaves the sector capital generally leaves as well. The drain on equity capital from the farm sector has been quite sub-stantial. This has led to greater use of debt capital in the sector than would otherwise have been the case. The result is a structure which is less resilient in the face of decreasing returns than would be the case where a larger share of capital was available in the form of equity.

Adjustment Through Each Individual's Life Cycle

Further adjustments are on-going for each individual farmer. Young, beginning farmers generally use substantial debt. As they accumulate equity through retained earnings they may expand rapidly through their middle years continuing to utilize debt financing. Again, they may use debt to consolidate a viable sized family farm which has been divided among several children or to expand to keep pace with technology. In later years, they may retire debt, scale the operation back in size, and invest equity in financial assets. In other cases, farmers may use debt to expand their operations enough to be able to bring sons into their businesses. Here, their goal will be to achieve a size which can support two operators. All in all, there will normally be a great disparity in the distribution of debt financing. At any historical moment, debt conditions will depend on what stage of the financial life-cycle the farmer finds himself in.

If the bust happens when any of these people have just contracted a heavy debt load, they will suffer. Some will have to leave farming. We think it important to note the arbitrariness of all that. Booms and busts are somewhat arbitrary. When the situation is ripe for a bust, say, the triggering event may be external to farming--a tax change, a shift in monetary policy, an alteration in relations with a foreign nation. If the conditions which will precipitate the bust hold off for five years, then a number of farmers will develop the strength to weather the storm. But then it will create trouble for others. No matter when a bust occurs, people will suffer. But both the when and the who are variable--we think arbitrarily so. As a result, trying to fix blame, as some government officials have done, seems misguided and pointless.

Implication for Farm Problems

We have said that greater debt causes the farm sector to be less resilient in the face of lower than expected returns. Our point here is that certain segments of the sector will quite naturally be more susceptible to periods of lower than expected returns. Our system of financing American agriculture guarantees that there will at all times be a number of individuals with relatively high debt. Thus our system assures a less resilient group at all times.

28

NOTES

1. This figure does not show a curve in which Miller included a return to land (Miller, _et al_., viewed land as the limiting resource, others have used such factors as operator labor or equity). That curve, which included all costs, by construct, asymotically approached from above a unit cost of $1.

3

The Current Situation of the Farm Sector

Net Farm Income

As we try to develop an understanding of the financial status of the farm sector in the years since 1950,
net farm income appears to be an important measure. It
seems to convey a final word about financial health. Real
net farm income for the sector--that is, net income
adjusted to allow for inflation and variations in the
value of the dollar--fell rapidly through the mid 1950s,
continued to fall but more slowly during the 1960s, and
skyrocketed during the early 1970s. Then it fell again,
even more steeply than it had risen, until, at present, it
is at the lowest level since the Great Depression--when,
in 1932, it went as low as $2.02 billion, in 1985 dollars.
The solid line in Figure 3.1 details that story.

However, net farm income for the sector can create a
mistaken impression about the financial health of farms.
As Figure 3.2 shows, the number of farms decreased from
roughly 6 million to only about 2.3 million between 1950
and 1983 as farmers substituted capital for labor and consolidated farms. Real capital consumption, or depreciation, per farm grew at an annual rate greater than 10 percent during that period. At the same time capital consumption expanded, the number of farms decreased rapidly--
at a 2.7 percent annual rate, so that real net farm income
per farm (see the dashed line in Figure 3.1) grew moderately from the mid 1950s until the boom period. Since
1973, real net farm income has fallen steeply, even on a
per farm basis. The important point to see in all of this
is that the current level of net income per farm more
nearly resembles that of the mid 1950s than it does that
of the depression years as the sector measure suggests.

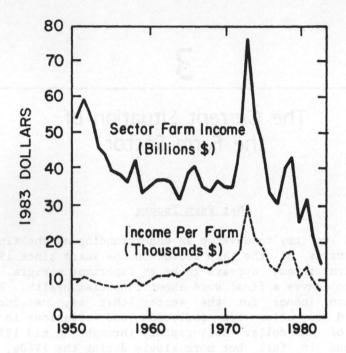

Figure 3.1.   Real Net Farm Income.

The per farm picture, though not rosy, is less bleak than
the sector picture.

New Wealth

On balance, a look at the real capital gains, sector
wide, between 1972 and 1984, suggests a very bright picture.   As a whole, the sector has amassed substantial new
wealth.   From 1972 through 1979, the actual boom period,
real capital gains added about $447 billion of new wealth
to the agricultural economy--in terms of 1983 dollars.
Even though the sector experienced real capital losses
totaling $149 billion between 1979 and the beginning of
1984, there remained almost $300 billion of new wealth.
But along with the new wealth came many billions of dollars of new debts as farm owners transferred assets at
higher prices (Melichar, January 1984, pp. 5-6).

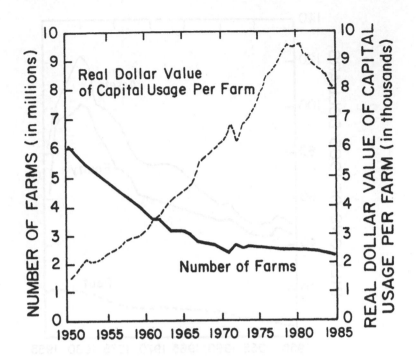

Figure 3.2. Capital Consumption (Depreciation) Per Farm and Number of Farms.

Figure 3.3 shows the patterns of change in real farm assets, debt, and equity throughout this period. When we consider the entire time period, the changes are dramatic. Also, the measures of assets and equity seem to follow three analogous trends. Between 1950 and 1971, sector wise assets and equity increase gradually. From 1972 to 1980, they increase sharply. And since 1981, they have fallen almost as sharply as they had risen during the previous eight years. However, even though real equity in the farm sector has declined about 40 percent from the peak level of 1980, it remains about 44 percent greater than it was before the boom in 1969. Notice, too, that the average real equity per farm has increased substantially--from $227,000 in 1969 to $271,000 in 1985 (Figure 3.4). That is a 20 percent increase.

Interestingly, the plots of equity, assets, and debt for sector and individual farms resemble each other closely. Both give the general impression that,

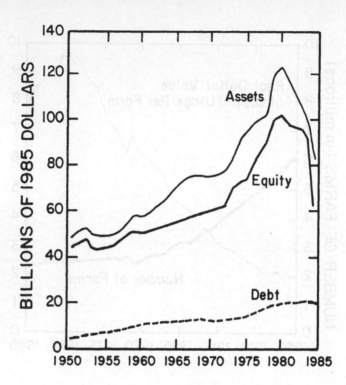

Figure 3.3. Real Assets, Debt and Equity for the Farm Sector.

throughout the sector, this has been a period during which farm wealth has increased substantially.

Profitability

Besides adding significant new wealth, the sector appears to have improved its profitability. However, that claim requires caution. Economists can choose among several approaches to measuring profitability. And, as Davis has said, they must be careful to distinguish between income problems related to the on-going farm size adjustment process and those attributable to the disruptive effects of the recent land boom-bust:

A serious error in reasoning about costs is introduced by the handling of capital charges, including interest, at prevailing rates for time loans on operators' investment. These costs, which contribute most heavily to the unfavorable conclusion reached

34

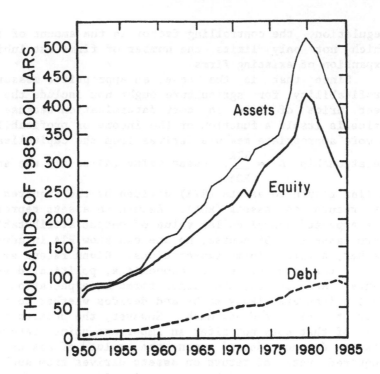

Figure 3.4. Real Assets, Debt and Equity Per Farm.

[low farm income] are calculated not on the actual
investment by farm operators, for which no data can
be had, but on land values presumably containing substantial increments of appreciation over and above
actual investment. The result is an inflation of
costs (Davis (1939), p. 72).

Davis compares this to public utility rates. If
these rates are allowed to rise, the profitability of the
industry is increased. If utilities were a "normal"
industry, this would provide an incentive for new producers to enter the industry, which would lower rates and
restore profits to normal levels. But since utilities are
regulated monopolies, this will not occur and the value of
existing fixed assets will rise to a level consistent with
a normal rate of return. It would clearly be fallacious
to use this value in computing costs for the purpose of
justifying higher rates. An analogous situation exists
for agriculture. The only difference is that instead of

35

regulation, the controlling factor is the amount of land, which not only limits the number of firms but inhibits expansion of existing firms.

Since that is the case, an appropriate measure of profitability for agriculture ought not include the current price of land in cost determination because that price is itself a function of the income or profitability. A more appropriate measure derives from the capitalization relationship $A = \dfrac{RTA}{ROA}$. Asset value (A) equals the annual dollar return to assets (RTA) divided by the required rate of return on assets (ROA). Return to assets represents the expected annual dollar value of output attributable to farm assets. Of course, no one can know RTA in advance. Rather, a farmer forms expectations. Given recent experience, the projections of economists, political developments, and so on, the farmer forms an expectation about what prices are likely to be and decides whether to plant, what to plant and how much. However, the uncertainty of all of that does not alter an important point--returns to assets do not depend upon the value of assets. The required rate of return on assets derives from such factors as expected risk, alternative investment opportunities, and the cost of borrowed funds. Finally, the market determines asset values according to the relationship between RTA and ROA. Given an expected level of return to assets and a required return on assets, asset values adjust to obtain equilibrium. Accordingly, dollar return to assets represents a measure of the profitability of agriculture which is not muddled by the value of agricultural assets.

Melichar measures the profitability of agriculture as the flow of real income from farm assets. Here the flow of real income is the same as what we refer to as return to assets. Crucial to his analysis is an estimate of profit derived from the utilization of farm assets. Figure 3.5 plots Melichar's thirty-year series for real income from farm assets as the solid line. Melichar deducts nonfactor operating expenses, depreciation, and imputed charges for labor and management from gross income, before interest payments on debt.[1] Measuring income in this way avoids confounding the issue of the levels of real returns to farm assets with problems attributable to excess debt. For the same considerations, he does not include capital gains, or losses, in this measure of profitability. Income returns denote realized

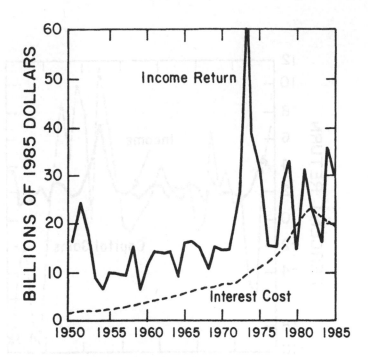

Figure 3.5. Income from Assets: A Measure of Profitability in Agriculture.

cash flows, whereas the total returns to ownership and control of assets include unrealized changes in value which derive in part from returns and expectation for future levels of that return. Melichar finds the profitability of agriculture by this measure to be above pre-boom levels, as the solid line in Figure 3.5 shows. Real income to farm assets averaged $27 billion from 1979 through 1985, or 93 percent above the average for 1960 through 1969, which was $14.2 billion 1985 dollars.

We have seen that real income returns figure significantly in determining the value of farm assets. Moreover, we can see that if we can expect real income returns to remain at higher levels, the same assets will have greater values. Other things equal, asset values tend toward the level where the income rate of return on assets is constant. The solid line in Figure 3.6 displays the rate of return on assets and shows it to fluctuate around 2 percent. Accompanying the increase in dollar returns in 1972, real capital gains, the dashed line in Figure 3.5, accrued as asset values increased. The higher asset

37

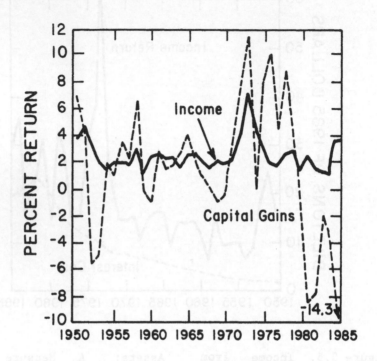

Figure 3.6. Components of Total Return on Assets: Income and Real Capital Gains.

values acted to bring the income rate of return back toward its pre-boom level of about 2 percent by 1976. Further, changes in asset values continued in a manner to maintain that approximate rate of 2 percent.

Changes in Value of Assets and Liabilities

While income return on assets has tended toward a constant rate, capital gains and losses have been far from constant. These substantial gains and losses serve as an adjustment mechanism which brings about the corrections in the other variables. Expressed as a percentage rate, these gains and losses are quite volatile. The dashed line in Figure 3.7 shows total returns on assets--that is, income returns plus real capital gains.

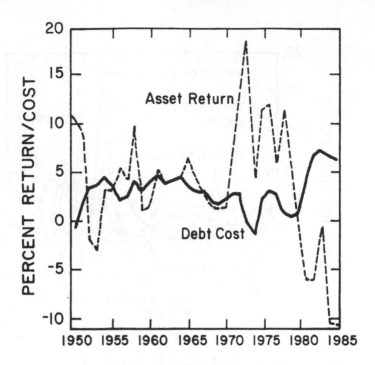

Figure 3.7.   Real   Total Return on Assets and Real Cost of
Debt.

     Not   an isolated factor, capital gains interact
closely with the real cost of debt. Accordingly, Figure
3.7 plots both. Real cost of debt represents the sum of
the nominal interest rate on the debt and the real capital
losses, or gains, on the debt. Given a fixed nominal
interest rate, inflation acts to decrease the value of
liabilities as well as the opportunity cost of investments.   Therefore, the real cost of debt should be
inversely related to real capital gains and so to the
total return on assets. Even though the causal relationships and time lags are complex and difficult to assess at
best, such an inverse relationship appears significant
during the past three years. Figure 3.8 plots realized,
unrealized, and total returns together with like measures
for debt costs and the return on equity to show what a
problem this is. Yet it seems possible to conclude here
that, in general, agriculture has been more profitable
after the boom period than it was prior to 1972.

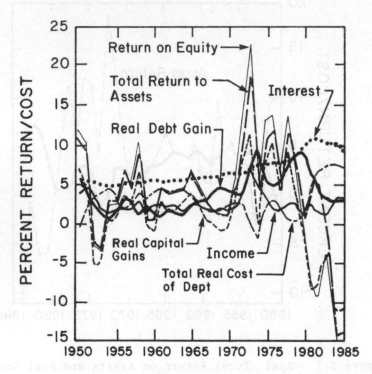

Figure 3.8. Complex Relationships Between Realized (Income), Unrealized (Capital Gain), and Total (the Sum) Asset Returns and Debt Costs.

Cost of Debt

Working against that happy conclusion, to some extent, are the facts that both debt, shown in Figure 3.3, and real cost of debt, the solid line in Figure 3.7, are higher now than before the boom period. Interest payment on debt, the dashed line in Figure 3.5, is more than four times greater than it was during the mid 1960s. That payment exceeds the increase in income return and leaves the sector with fewer funds available for investment, consumption, and savings than it had before the boom. In fact, for the sector, return to assets fell short in meeting interest on debt by a total of $15 billion 1985 dollars over the four year period 1980-83 (Melichar, July 1984, p. 4). Inability to meet interest obligations out of current income is clearly an important aspect of the

40

present financial situation that distresses so many farmers.

Summary of the Current Situation

Looking at any one of these factors--net farm income, profitability, changes in asset values, real cost of debt --can distort our view of the farm financial situation. Some, taken alone, motivate optimism. Some, pessimism. A broader, sector-wide consideration, also, may produce different conclusions from analyses of individual farms. A comprehensive view of all these measures, on the other hand, provides a useful general picture, one which calls attention to important problem areas--cash flow and asset valuation.

The combination of new wealth and higher real income to assets, compared to pre-boom levels, makes farmers with very low, or no, debt better off. However, the remaining real income return to assets after interest payment has fallen to levels below those before the boom. After interest payments, the sector as a whole is worse off. To consider what is happening to many farmers, consider a representative case. Assume a farmer with a debt-asset ratio of 24 percent, who pays 9.8 percent interest and earns income based on a 2.4 percent rate of return on assets. That farmer realizes about $2.40 income return for each $100 of assets he controls and pays out $2.36 in interest on debt. His realized return on equity is essentially zero.

That hypothetical farmer may be better off than many. Nationally, the average farm equity is $271,000. Given that, the average farmer must balance an average return to labor and management of $8,000 and average non-farm income of $18,000 to cover the needs of family consumption, net investment, savings, and funds for buying the equity owned by people who are leaving the sector.

The greater the debt and the higher the interest, the worse off the farmer--unless he is earning an exceptionally high rate of return on assets. But among large and medium sized farms, perhaps one-third have great enough debt that they are under severe financial stress--that is, they are losing equity at rates which make their chances for survival an issue of national concern.

NOTES

1. Non-factor operating expenses include feed, purchased livestock, seed, fertilizer and chemicals, fuel,
repairs, and machinery expenses as well as hired machine
work, business taxes, and other miscellaneous expenses.

4

Financial Conditions of
Different Sizes and Types
of Farms

Those responsible for farm sector policies always take farm financial conditions into consideration during the policy formulation process. Now, though, the severe financial stress and the associated bankruptcy problems have created far broader awareness of farm finance among people in all walks of life and have made farm economics a pressing public issue. Policy makers must realize, however, that there are degrees of financial stress, of which insolvency leading to bankruptcy is only the most extreme. Further, since a sound, comprehensive policy should speak to the needs of those all along the spectrum of financial stress, policy analysts could profit from having a more comprehensive description of farm financial conditions.

The financial conditions of farms vary along several dimensions. First, analysts can order standard financial variables from those causing least stress to those causing most stress and then classify farmers, in any given year, according to those categories of stress. Secondly, analysts can identify reasons for experiencing stress in a given year. One reason surely relates to a farmer's degree of financial leverage. An important negative aspect of leveraging is its tendency to magnify the effects of the business variables. That is, if commodity prices are weak, a highly leveraged farmer will suffer far more than one who carries less debt. Also, a farmer's level of managerial skill will effect the degree to which fluctuations among the variables induce stress. Thirdly, in a given year, a certain kind of farm, or a certain farming region, may experience greater financial stress than other kinds or places. Finally, the size-profit relationship creates another systematic variation in stress.

A taxonomic system of that kind can account for the complexity of the farm financial situation. It allows students of farm policy to rank farms on a scale of financial stress both for individuals and for the entire sector.

Conceptual Issues

The notion cash flow problem, which agricultural economists typically use in accounting for the problems of financially stressed farmers, may obscure more than it reveals. To begin with, cash flow problems may symptomize several conceptually different underlying problems. Also, those who use cash flow as a key notion have yet to develop comprehensive categories of stress. In our view, the notion profitability, which subsequently relates to cash flow, provides a far more useful measure of the situation.

In our discussion, the asset structure of a farm business includes depreciable assets and land. The distinguishing features of these assets ultimately require further discussion. Also, to us, a typical farm business approach to financing includes debt on a permanent basis rather than holding to the long term goal of reducing debt to zero. Thus, for profitable farms, this implies expansion--an attractive view for two reasons. Further, the continuing process of growth and adjustment to technological change provides incentives for expansion of real dollar asset holdings in the sector. The declining number of farmers implies greater asset holdings for each farmer. These underlying notions provide a foundation for a definition of an individual farm which will feed a useful analysis of the degrees of farm stress. However, it also allows incorporation of phenomena in this consideration best described through sector level observation.

There is an important relationship between profitability (in the net present value sense) and cash flow. That is, an investment with a positive net present value will more than pay for itself (including interest) if payments are accepted according to cash flows from the investment. Conversely, investments that are not profitable will not pay for themselves. Thus, it is of first and foremost importance to know if farms are profitable.

However, several factors can complicate the repayment of profitable debt financed investment, even without considering variations in returns over time. For one, lenders may be impatient about waiting for cash flows from investments with relatively small operating cash flows and large capital gains. For another, lenders may require very fast payoff for depreciating assets to insure collateral sufficiently greater than principal throughout the life of the loan. These problems relate to both land and machinery. The nature of the relationship between income return to land and land value generally results in a cash return below the debt service requirements of lenders (Dobbins, et al., 1981). And the gap between the flow of income and debt service requirements widens with inflation in the case of level payment loans. Also, heavily debt financed machinery purchases often accompany rapid farm expansion, even when the expansion involves rented acreage. Servicing this debt at the rate preferred by lenders may well be difficult.

The financial stress of mismatched returns to assets and debt servicing is real and the cause of agony for farmers throughout history. However, this kind of cash flow problem is not as serious as one resulting from nonprofitability. When the schedule of income does not match the payment schedule but the farm is basically profitable, restructuring the debt can alleviate the pain on all sides. No amount of debt restructuring can solve the problem of the unprofitable farmer.

As central a notion as profitability is in this kind of discussion, it presents analysts with a variety of problems. We have trouble measuring profit accurately even when we have relatively complete accounting data. Worse, the very nature of the problem is complex and defies simple approaches.

For example, it seems that, in any given year, some farmers will be unprofitable because of the business risk involved. That is, weather patterns, prices, equipment trouble, labor problems--any or all of those factors create a dynamic situation from which some will benefit and which others will find detrimental to profitability. Perhaps there is an analogue here to the full employment notion which holds that even during a time of nominally full employment a certain number of people will not find work. Perhaps, that is, a normal percentage of farms will be unprofitable each year, no matter the general condition of the farm economy. If such a percentage of unprofitability were found to be relatively constant, then that figure would enter significantly into any policy discussions about the farm economy.

Also, policy makers could no doubt benefit from having this kind of information, along with other kinds, across farm sizes, kinds of farms, and geographic locations of farms. For example, if 10 percent of midwest crop-livestock farmers are unprofitable this year, what percentage of that group also lost money last year; and very importantly, what is the probability of a farmer losing money in both years?

Under the best of circumstances, measuring profitability is difficult. The sense of profitability implicit throughout our discussion has its basis in the notion of net present value--and we assume that annual accounting data of the usual kind will suffice as we measure historical profitability outcomes. Estimates of net present value for a proposed investment typically include charges for all resources including the owner's equity capital--at the marginal re-investment rate for equity, the opportunity cost--and the owner's labor and management efforts--at the opportunity cost or reservation wage, whichever is higher. However, separating the returns to unpaid labor, management, and equity is impossible without making assumptions about the accounting data. Economists and investors seldom test these assumptions. It may be that they are not testable.

Thus, we end up with the flow of returns to the human resource (labor and management) instead of separated which would be more useful. This in turn requires us to deal with the use of funds for consumption as well as for investment and debt repayment.

Another factor complicating the accurate use of annual accounting data to measure financial stress is the handling of asset values during the interval between acquisition and disposition. A net present value model includes cash flows over the whole investment horizon. For land, a major cash flow is the expected gain, or loss, at the end of the planning horizon. A farmer will not realize this result until the sale of the asset and so it does not appear in the cost basis accounting data because that includes only realized results. Such balance sheet data are available for the agricultural sector in toto but only sporadically available for individual farms. In spite of that limitation, we can use beginning and ending market value balance sheets to compute the change in unrealized gains and losses during a given year, and then treat that result like income.[1] As Melichar says, it is

often useful to remove capital gains from the income measure for policy analysis--because the capital gain itself
is a function of changes in the current return to land.
However, capital losses or gains provide an important measure of stress for an individual farm, so we need to be
able to identify them accurately. As Melichar says, it is
often useful to remove capital gains from the income measure for policy analysis because the capital gain itself
is a function of changes in the current return to land.
However, capital losses or gains provide an important measure of stress for an individual farm, so we need to be
able to identify them accurately.

Accounting Measures

Certain traditional accounting relations among income
cash flow and changes in net worth enter into the definition of profitability in terms of net present value.
Table 4.1 outlines the crucial aspects of these accounting
relationships. However, various assumptions concerning
these relationships and what they may imply about a farmer's financial condition and the interplay among these
relationships require careful attention lest they be mis-
leadingly construed. We take net income to be the realized farm operating returns to the owner's labor, management, and equity capital.[2]

If net income is positive, then the farmer has earned
sufficient income to cover his non-factor operating costs
(see footnote 1), interest, and to replace his capital in
an amount equal to depreciation.

More specifically, this is what the relationship in
Table 4.1 expresses. "Net investment, the second relationship, specifies the extent to which expenditures for
capital items exceed capital consumption.[3] Zero level net
investment generally implies that the farm size is holding
constant, but since most farms have depreciable assets,
this does not imply a complete absence of capital
expenditures.

A positive level of retained earnings (item 4, Table
4.1) implies that combined farm and off-farm incomes are
more than sufficient to cover income taxes and family living expenses.[4] A farmer feels some financial stress when
retained earnings are negative. In this situation he must

47

Table 4.1. Accounting Relationships.

(1) Farm Income[a]

+ Gross Farm Income
− Non-factor Operating Costs
− Interest
− Capital Consumption
= Net Farm Income

(2) Net Investment[b]

+ Capital Expenditure
− Capital Consumption
+ Additions to Inventory
− Reductions in Inventory
= Net Investment

(3) Debt

+ Additions to Debt
− Reductions in Debt
= Net Borrowing

(5) Sources of Funds[d]

+ Retained Earnings
+ Gifts and Inheritances Less Gifts Made
+ Net Borrowing
= Total Sources of Funds

(6) Uses of Funds[d]

+ Net Investment
+ [Funds for Farm Transfers][f]
= Total Uses of Funds

(7) Change in Market Value Networth

+ Retained Earnings
+ Gifts and Inheritances Less Gifts Made
− [Funds for Farm Transfers]
+ Increase in Unrealized
  Capital Gains[g]
= Change in Market Value Net Worth

48

Table 4.1 (continued).

(4) Retained Earnings[c]	(8) Change in Real Net Worth
+ Net Farm Income	+ Change in Real Cost Basis Net Worth
+ Off-farm Income	− [Real Net Funds for Farm Transfers]
− Family Consumption and Income	+ [Real Gifts and Inheritances Less Gifts Made]
− Taxes	+ Real Gains on Assets During Year
= Retained Earnings	+ Real Gains on Debt During Year
	= Change in Real Net Worth

a If data from cost basis accounts are available inventory change should be included to properly account for the cost of goods sold in the expenses.

b If cost basis data are available this category would not include inventory changes and would include a subtraction of the cost basis of assets sold.

c When data is available (especially in the case of individual farmers) this section should include realized capital gains.

d There are many types of sources and uses of funds statements. This one is based upon the statement of changes in financial condition.

e At the sector level this would net to zero.

f Applicable for the sector aggregate only and is in essence the purchase by remaining farmers of the equity in agricultural assets of discontinuing proprietors.

g This can be negative.

cover living expenses through borrowing or through negative net investment ("living off of depreciation"). Items 5 and 6 in Table 4.1 show this.

In order to be able to apply the sources and uses of funds relationships to the agricultural sector as a whole, we must consider one additional use for funds--farm transfers. People leaving the sector such as retirees, young adults who choose not to farm, or heirs who cannot for some reason farm--transfer ownership of some of their farm assets to remaining farmers. Subsequently, they invest the cash which they receive for their equity in those assets in other sectors of the economy. Similarly, new entrants into the sector may bring to it new equity. The balance--that is, the net funds for farm transfers, represents a substantial cash drain from the agricultural sector, and is a factor which contributes substantially in determining total agricultural sector debt. Note that this account is not applicable for an individual farmer who would gain a new asset if he purchased another's asset.

The degree of financial stress caused by negative retained earnings depends upon other factors--primarily the farmer's credit reserve and changes in unrealized capital gains. A farmer with substantial credit reserve can withstand negative retained earnings for a number of years before being forced to liquidate his business. At this point, we can see how important a factor capital gains can be. While this source of unrealized income can be positive or negative in a given year, we normally expect a general trend of positive nominal capital gains as the general price level in the economy increases.[5] An increase in unrealized capital gains could offset negative retained earnings and prevent net worth from falling (Item 7, Table 4.1). We might logically expect net worth to increase at least at the rate of general prices to maintain the buying power of a farmer's wealth. Thus, to eliminate money illusion from the measure one should measure the change in real net worth.[6]

Under circumstances of negative retained earnings, a constant or growing real net worth would imply that normal unrealized capital gains were sufficient to offset the negative retained earnings and to maintain the farmer's buying power at the level of his net worth. This situation should pose no severe financial difficulty. However, negative real net worth change would in all likelihood erode a farmer's credit reserve and eventually lead to a major cash flow problem.

These accounting relationships define a range of financial situations--a variety of ways in which a farmer can be profitable and a variety of ways in which he may suffer financial stress. Accordingly, this way of looking at accounting information provides a framework for our understanding of a series of levels of financial stress which farmers may experience.

Categories of Stress

In defining categories of stress, we focus on four critical financial variables--net farm income, retained earnings, change in nominal net worth, and change in real net worth. For each variable, zero dollars is the critical level for measuring stress.[7] Clearly, a positive net farm income differs qualitatively from a negative one. And so it is with each of the variables.

Then, if we assume that each variable can have either a positive or a negative value, we can define sixteen financial situations, as Figure 4.1 shows. However, in practice, we think eight categories sufficient for useful discussion. One of these identifies a prosperous farm situation, the other seven--degrees of stress.

To arrive at that conclusion, we first recognize that the order of these variables is significant. Net farm income is an important measure of profitability. Obviously, a financially prosperous business has some level of positive profit. A farmer with a positive net farm income is covering variable expenses and replacing depreciable capital. Also, he has some funds for consumption, and if retained earnings are positive, he can finance some expansion with equity rather than debt financing. Even when retained earnings are negative, increasing nominal net worth can relieve the financial stress that results. Finally, change in real net worth provides an extremely sound measure of a farmer's condition. For when this is positive, it can alleviate, in some degree, negative values for prior variables. But when this value is negative, it can overwhelm positive values for prior variables.

To simplify the stress taxonomy, we have eliminated several of the 16 possible outcomes which Figure 4.1 specifies. A positive change in real net worth following a

51

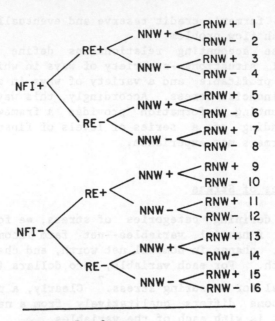

```
                                      RNW +  1
                         NNW + <
                                      RNW -  2
                RE+ <
                                      RNW +  3
                         NNW - <
                                      RNW -  4
      NFI+ <
                                      RNW +  5
                         NNW + <
                                      RNW -  6
                RE- <
                                      RNW +  7
                         NNW - <
                                      RNW -  8


                                      RNW +  9
                         NNW + <
                                      RNW - 10
                RE+ <
                                      RNW+  11
                         NNW - <
                                      RNW- 12
      NFI- <
                                      RNW + 13
                         NNW + <
                                      RNW - 14
                RE- <
                                      RNW+ 15
                         NNW - <
                                      RNW- 16
```

NFI = Net Farm Income
RE = Retained Earnings
NNW = Nominal Net Worth Change
RNW = Real Net Worth Change

Figure 4.1. Various Financial Outcomes.

negative change in nominal net worth is possible only
under deflation. Because of the unlikely nature of this
result, we select sequences 3, 7, 11, and 15. Also, we
regard the outcomes revolving around a positive retained
earnings following negative net farm incomes as rather
special cases and treat them as subcategories. As we progress from no stress to dire stress, the first four categories describe situations which show positive growth in
real net worth. Table 4.2 provides an outline of the
eight categories we find meaningful.

Category 1 (No Stress)

The first category suggests a set of farmers who are
not experiencing stress. Not only has their real wealth
increased but growth has been partially equity financed,
or if no growth has occurred the level of debt has been
reduced. Farmers expect to be in this category most
years. Economists might discuss the level of returns

Table 4.2. Categories of Financial Stress.

Category	Financial Outcomes
1.	Positive growth in real net worth Positive retained earnings Positive net farm income
2.	Positive growth in real net worth Negative retained earnings Positive net farm income
3.	Positive growth in real net worth Positive retained earnings Negative net farm income
4.	Positive growth in real net worth Negative retained earnings Negative net farm income
5.	Negative growth in real net worth Positive retained earnings Positive net farm income a. positive growth in nominal net worth b. negative growth in nominal net worth
6.	Negative growth in real net worth Negative retained earnings Positive net farm income a. positive growth in nominal net worth b. negative growth in nominal net worth
7.	Negative growth in real net worth Negative retained earnings Negative net farm income Positive growth in nominal net worth
8. (Most Stress)	Negative growth in real net worth Negative retained earnings Negative net farm income Negative growth in nominal net worth

relative to the opportunities, the degree of profitability. Whatever the details of the particular case, though, these situations should not involve significant financial stress.

Category 2 (Slight Cash Flow Stress)

Here current income fails to cover family withdrawals, so farmers must increase debt to cover consumption and to finance growth. Farmers in this category may perceive the need to use additional debt financing and feel a small amount of cash flow stress. Their credit availability should be expanding at least as rapidly as borrowing since their real net worth is increasing.

This situation can occur when inflation is relatively high, with real net worth increasing due to increasing real returns to assets. A leveraged farmer will be subtracting high nominal interest from current income, so for him a trade-off between current income and capital gains occurs. This situation would be common in situations of relatively high inflation with generally good farm sector conditions and is more likely to effect those with higher leverage.

Category 3 (Current Farm Income Stress Alleviated by Good Off-Farm Income)

During an off-year during the boom part of a cycle, or at the end of a boom period, some farmers might experience mild stress. Because of positive growth in real net worth, though, most people will consider general farm sector prospects to be favorable. That, along with asset value increases will lead to expansion of credit availability. Farmers in this category will have concern about the level of current farm income, but a substantial off-farm income will offset negative farm income and amply cover withdrawals. Accordingly these farmers will experience little cash flow stress. It is probably the case that few commercial farmers experience this situation. Rather, most of those in this category are hobby or tax loss farmers--with the notable exception of those who consider themselves commercial but are striving to build to a viable size.

Category 4 (Modest Income and Cash Flow Stress)

This situation might occur for farmers who have a bad year when the agricultural economy on the whole is good. For example, some farmers might be hit especially hard by weather. Some percentage of the least skilled managers might also experience this kind of stress on a regular basis. Activity in this category may have predictive power. A marked increase in the number of farmers finding themselves in this category may signal the end of a boom cycle.

This category involves fairly significant cash flow stress within a general agricultural situation favorable enough to produce real asset value increases. Farmers in this situation are financing family living with debt or are living "off of depreciation."

Category 5 (An Illusion of Well-Being)

While economists tend to discount money illusion, most farmers in this category probably do not realize the severity of their situation. Several attributes of this situation may mislead them to then suffer from an illusion of well-being and fail to take action to protect themselves from further stress. That, by any other name, is money illusion. If nominal net worth has increased (5a) there is little or no cash flow stress. That creates a false sense of security, for it obscures the fact that the farmer's real wealth is decreasing--a fundamentally undesirable situation. Many well-managed farms would be in this situation in the early part of a general downturn in agriculture if the inflation rate is high enough.

If a farmer's nominal net worth has declined, though, he will perceive the situation to be substantially worse. If the general situation in agriculture is severe enough for nominal asset values to decline along with everything else, then the farmer will need well above average management skill even to maintain a significant amount of leverage and remain in this category and not get into a worse situation. Also, farmers with no debt might be in this category under conditions where land values are increasing at a rate slower than the rate of inflation (5a) or where there is a decrease in nominal land values (5b). Their situation resembles, in some degree, that of the farmers whose real wealth is eroding.

Category 6 (Financially Unsound)

This category covers the same economic conditions as Category 5 except that retained earnings are negative. This would induce significant cash flow stress for farmers without substantial credit reserves, because they lack the growth in real net worth which would support expanded credit availability.

Category 7 (Major Stress)

Category 7 combines major cash flow stress with a near disastrous income and wealth situation. An unleveraged farmer is unlikely to have negative net farm income due to situations other than bad luck (such as bad yield) or unusually poor management. So most of the farmers in this category are probably over-leveraged for the economic conditions of the moment. The only positive factor in this situation is a growth in nominal net worth, probably an accident of high inflation.

Category 8 (Certain Road To Bankruptcy)

There is little hope unless being in the category is a purely random, one-time occurrence. In this category survival depends upon the initial magnitudes of net worth --how long before it is gone. In the bust part of a land boom a number of farmers with high leverage will find themselves in this category.

These eight levels are generalizations, of course, though they appear to hold in a vast majority of cases. However, the complexity of the economic situation may influence the degree of stress a farmer experiences. Even farmers who appear, given most accounting measures, to be free of stress may experience some. And there are cases where leverage, usually a contributor to stress, may actually alleviate it. For example, consider the interplay of interest rates and current returns on assets. When economic conditions or other events specific to a firm result in a current or operating rate of return on assets less than the interest rate (the common situation when averaging across farms), net farm income, and therefore retained earnings, will decrease with financial leverage. Thus, a farmer with greater leverage may have the same current or operating rate of return on assets but be in greater financial stress. This correlation between stress and leverage permeates the stress categories. For the relatively small portion of farmers with an operating rate of return greater than the interest rate they pay, leverage would reduce their financial stress.

The discussion above has focused upon return from operating the business. The balance of returns to a farm business are in the form of unrealized capital gains or losses. Unrealized capital gains or losses are self-financing and do not directly involve cash flow. However, there is an extremely important indirect link to cash flow via a business' credit reserve. Thus, we have another link between credit and financial stress.

The primary asset with substantial capital gains or losses is real estate. Thus, the boom-bust cycle in real estate is tied into cash flow through credit reserve changes, as well as having a rather astounding influence on real wealth. The final linkage, of course, is the flow of residual operating return to assets (not to be confused with net farm income). Changes in this flow, and more importantly, the perceived future of this flow is the primary force driving real estate values.

Measuring Financial Stress in the Farm Sector

Financial information concerning the farm sector as a whole appears in some cases to counter information about individual farms. In other cases, there seem to be parallels. Finally, though, a weighing of general against particular can prevent our being misled by special cases and sharpen our insight into the entire problem of financial stress. For one thing, study of sector data as well as individual farm data can help us gauge how many farmers have cash flow problems and how severe their problems are. For another, these analyses call attention to what we regard to be the crucial focus of the problem--changes in asset values.

Throughout the last 23 years, retained earnings for the sector have remained positive, though they have fluctuated between $7.5 and $27.77 billion--as is also the case with net farm incomes. Movements in those two sectoral measures roughly indicate the number of farmers experiencing cash flow problems and also the severity of the problems.

Table 4.3 develops the retained earnings and net farm income situations, in nominal or current dollars, for the

Table 4.3. Derivation of Retained Earnings for the Sector, 1960-1983.
(Millions of Current Dollars)

Year	Gross Farm Income[1] (+)	Non-factor Operating Costs[1] (−)	Capital Consumption[1] (−)	Interest on Debt (−)	Net Farm Income (=)	Off-Farm Income[2] (+)	Consumption[3] (−)	Retained Earnings
1960	37054	20067	1268	3773	11946	8482	12920	7508
1961	38608	20864	1347	3802	12595	9163	13108	8650
1962	40350	22161	1478	3915	12796	9904	14096	8604
1963	41273	22958	1654	4043	12618	11020	14151	9487
1964	40108	22756	1803	4201	11348	11637	13863	9122
1965	44304	23986	1986	4360	13972	12727	16834	9864
1966	48097	26152	2214	4626	15105	13882	18217	10770
1967	47993	27286	2460	4967	13280	14495	16651	11124
1968	49223	27917	2643	5348	13315	15466	17396	11384
1969	53577	29735	2902	5655	15285	16612	19354	12543
1970	55785	31477	3204	5890	15214	17617	19115	13716
1971	58902	33398	3377	6331	15796	19110	19471	15435
1972	67708	36176	3667	6721	21144	21265	25525	16884
1973	95070	45465	4433	7540	37632	24714	34576	27770
1974	93610	49846	5430	8928	29406	28135	32283	25258
1975	95432	51519	6075	10604	27234	23901	32213	18922
1976	96930	57337	7013	11794	20786	26681	30572	16895
1977	101634	60450	8146	13166	19872	26120	36884	9108
1978	119326	67217	9788	14348	27973	29704	43310	14367
1979	141196	79880	12534	16123	32659	35165	49198	18626
1980	139407	84815	15637	17805	21150	37568	42070	16648
1981	155155	87522	19118	19519	28996	39835	48446	20385
1982	149584	88784	20982	19806	20012	39415	43867	15560
1983	139350	85050	20800	19250	14250	40993	42662	12581
Coefficient of Variation	.601	.655	1.215	.744	.393	.644	.549	.453

1 Melichar, July 1984.

2 Economic Indicators Branch, ERS, USDA.

3 Derived as Median U.S. Family Income (Economic Report to the President)
x ratio of farm to non-farm income (Agricultural Statistics)
x number of farms.

58

sector. Again, all of these data require a cautious
approach. The figures for net farm income are positive.
Yet the nominal increases have been less than general
inflation would account for. In real terms, then, net
farm income for the sector has declined since 1960. However, off-farm income has offset much of the decline.

Figure 4.2, which plots real net farm income and
total farm income, provides an interesting insight into
the cash problems of farmers. The hashed area in Figure
4.2 represents an estimate of consumption expenditure for
the sector. Total income less consumption derives
retained earnings.

Pointing out that real retained earnings indicates
how much cash the sector has available for net investment
and for purchasing the equity of those who are leaving
farming underscores the importance of retained earnings as
an indicator of well-being. As the lowest line of the
graph shows, an unusually large amount of cash was available in 1973 and 1974. Since land is not an expandable
resource, the presence of that much cash undoubtedly drove
up land prices from 1973 on. That is, the amplitude of
the retained income in those years helped fuel the boom.
Then, notice, retained earnings fell steeply after the
high of 1973 to reach the lowest level in 23 years by
1983.

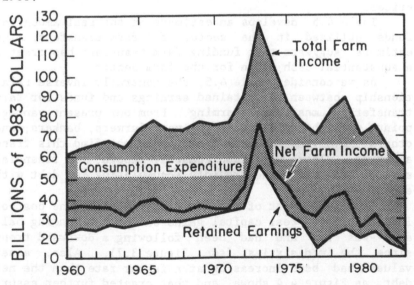

Figure 4.2. Real Retained Earnings: Net Farm Income +
Off-Farm Income Less Consumption

59

Trends in income, consumption, and retained earnings on a per farm basis differ from the sector measures because of the decreasing number of farms. Table 4.4 outlines the situation for the individual farm. Notice that both total farm income and consumption expenditures are higher now than during the early 1960s. In fact, Figure 4.3, which roughly parallels Figure 4.2, makes it appear that before the boom in the late '70s, consumption and income fluctuated more or less together but the greater increases in family consumption which the boom period allowed have not been adjusted for as total income has declined in the years since the 1973 peak. Apparently, then, patterns of family consumption are partly responsible for the low retained earnings we see at present.

Other things remaining equal, a decrease in real retained earnings implies that less funds are available for net investment or for farm transfer. However, increased real net borrowings partially offset the earnings decline between 1973 and 1979. Annual net borrowings for the sector peaked in 1979 at $31 billion (1983 dollars)--about five times the real pre-boom level. Real debt in the sector grew at a substantially greater rate during the 1970s than during the previous two decades (as Figure 4.3 shows), which helped to maintain net investment and to fund farm transfers, even at rapidly inflating prices.

Table 4.5 develops an estimate of the real amount of funds utilized in the sector for farm transfers. The estimate suggests that funding farm transfers has created a substantial cash drain for the farm sector.

As we consider Table 4.5, the generally inverse relationship between the retained earnings and funds for farm transfer columns seems alarming. From our present vantage point, we might well wonder why more farmers, bankers, and others involved in farm finance did not find this trend alarming. There were good reasons, though, for financial experts to view the rapidly accumulating new debt with equanimity.

The total cost of the new debt--that is, the cost of interest less real capital gains on the outstanding balance--was low and had been following a downward trend since the mid-sixties (see Figure 3.7). Also, asset values had been increasing at a faster rate than the new debt, as Figure 4.4 shows, and that created further assurance that all was well. As analysts observed the interplay of these indicators, their reactions were quite the opposite of alarm, for the ratio of debt to assets for the

Table 4.4. Derivation of Real Retained Earnings per Farm (1983 dollars).[a]

Year	Gross Farm Income +	Non-factor Operating Costs -	Capital Consumption -	Interest on Debt -	Net Farm Income =	Off-Farm Income +	Consumption -	Earnings =
1960	28941	15673	990	2947	9330	6625	10091	5864
1961	32517	17572	1134	3202	10608	7717	11040	7285
1962	33477	18386	1226	3248	10616	8217	11695	7138
1963	36728	20430	1472	3598	11228	9806	12593	8442
1964	36459	20686	1639	3819	10316	10578	12602	8292
1965	39531	21402	1772	3890	12467	11356	15021	8802
1966	43775	23802	2015	4210	13748	12635	16580	9802
1967	46882	26654	2403	4852	12972	14159	16266	10866
1968	46278	26246	2485	5028	12518	14540	16356	10703
1969	49177	27293	2664	5191	14030	15248	17765	11513
1970	52380	29556	3008	5530	14285	16542	17949	12878
1971	55499	31468	3182	5965	14883	18006	18346	14543
1972	55723	29772	3018	5531	17401	17501	21007	13895
1973	76378	36526	3561	6058	30233	19855	27778	22310
1974	67513	35950	3916	6439	21208	20291	23283	18216
1975	64471	34805	4104	7164	18398	16147	21762	12783
1976	63026	37282	4560	7669	13515	17348	19878	10986
1977	60819	36174	4875	7879	11892	15631	22072	5450
1978	70109	39493	5751	8430	16435	17452	25446	8441
1979	76306	43169	6774	8713	17650	19004	26588	10066
1980	68451	41645	7678	8742	10385	18446	20657	8174
1981	70083	39534	8636	8817	13097	17993	21883	9208
1982	64786	38453	9087	8578	8667	17071	18999	6739
1983	59047	36038	8814	8157	6038	17370	18077	5331

Coefficient of Variation

| | .425 | .459 | .903 | .514 | .383 | .519 | .407 | .496 |

[a] Table 3 divided by personal consumption expenditure deflator and by the number of farms (USDA, Agricultural Statistics).

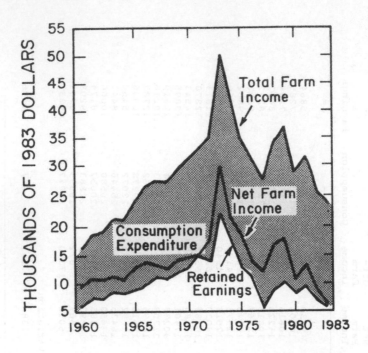

Figure 4.3. Real Total and Net Farm Income, Consumption and Retained Earnings on an Average Farm.

sector actually fell from its 1970 peak and held relatively constant through 1979.

What the analysts could not learn from these indicators was that major structural changes were occurring in
the farm economy. In real terms, 1980 net farm income for
the sector was the lowest in at least thirty years. To
understand just how low it was, we need only notice that a
26 percent improvement in 1981 still left that year the
third worst in three decades. In the most recent two
years, 1982 and 1983, as Figure 4.3 shows, net farm income
has reached further new lows, even on a per farm basis.

Retained earnings in 1981 were also very low; for the
average family, most of the benefit from recent gains in
off-farm earnings was absorbed by increases in expenditures for consumption. At the same time, tight monetary
policy and other changes in the economy initiated in late
1979 had begun to work through the economy to raise the
real cost of debt and dim expectations of continuing
inflation with its associated growth in asset earnings.
Asset values fell substantially in 1981, the first time

62

Table 4.5. An Estimate of Real Funds Used in the Farm
Sector for Farm Transfer Less Gifts and
Inheritances.

Year	Retained Earnings	Net Borrowing	Net Investment	Funds for Farm Transfer Less Gifts and Inheritances
	+	+	-	=

(millions of 1983 dollars)

Year	Retained Earnings	Net Borrowing	Net Investment	Funds for Farm Transfer Less Gifts and Inheritances
1960	22284	3265	2273	23275
1961	25424	4913	461	29874
1962	24913	7320	4859	27374
1963	27065	8120	4550	30633
1964	25648	7569	188	33028
1965	27268	10156	744	36680
1966	28946	9024	3935	34035
1967	29306	7790	8604	28490
1968	28717	4211	7277	25699
1969	30279	5905	3232	32951
1970	31642	5438	2113	34965
1971	34134	9827	7556	36404
1972	36029	11510	6827	40711
1973	56065	18072	15713	58423
1974	46306	14970	1349	59926
1975	32226	15651	12750	35127
1976	27376	17871	4166	41081
1977	13953	22089	10978	25062
1978	20563	22684	13598	29647
1979	24460	31365	14619	41205
1980	19847	18138	-4401	42388
1981	22412	16997	11791	27617
1982	16174	7092	2797	20468
1983	12581	2983	-23325	38889

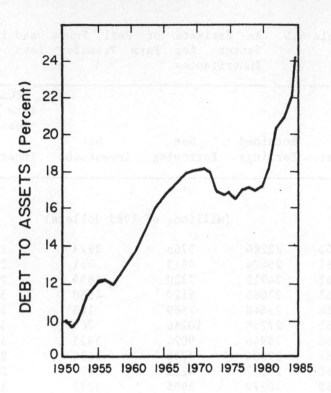

Figure 4.4.   Financial Leverage in the Farm Sector.

that had been the case since just after the Korean War
(1953-1954). Real capital losses on assets amounted to
8.6 percent. The real cost of debt, 3.9 percent, was the
highest since 1964 and rising. This was not the best time
to borrow. The years, 1982 and 1983, were no better. Net
farm income and, more importantly, retained earnings set
new lows. There was insufficient cash available for net
investment or for purchasing the equity of those leaving
the sector. Furthermore, during the increasing asset
values and declining real debt costs helped sustain credit
reserves and alleviate cash flow problems. Now, with
asset values plunging and debt cost soaring, there was no
easy way out. Worse, this complex problem caused many
farmers to increase the degree of financial leverage and,
simultaneously, to assume the added risk that invariably
accompanies greater leverage. Vast numbers of farmers had
to borrow just to meet rising interest costs on existing
debt obligations. As a result, the sector has experienced
a large increase in financial leverage since 1980 (Figure
4.4) and severe cash flow problems as individuals have

struggled to allocate limited, often negative, retained earnings among competing demands for their cash.

These considerations suggest that, while individual farmers may fit into any of eight categories, the sector as a whole has found itself in only three of those since 1950. Throughout that period, the sector has experienced positive net farm income and retained earnings. Thus, as Figure 4.5 reveals, the critical variables have been changed in nominal net worth and in real net worth.

During the entire period, nominal net worth has decreased during only two short periods--1953 through 1954 and 1981 through 1983, a total of five years. In addition, real net worth declined in only two other years than those--1960 and 1970.[8]

From that we can see that, as a whole, the farm sector has fit into Category 1 (See Table 4.2) during 80% of the years since 1950. Category 1 defines an absence of stress. The other seven years fit into either of the two subdivisions of Category 5, a situation marked by declines in real net worth. Until 1981 those declines were mild and followed by real equity growth. Although they fit into the same stress category, recent declines are more severe and future growth in returns seems unlikely in the near term.

An understanding of the severity of the current situation follows from consideration of statistics concerning real total return to assets, which include capital gains and losses and after interest payments. Figure 4.6 shows

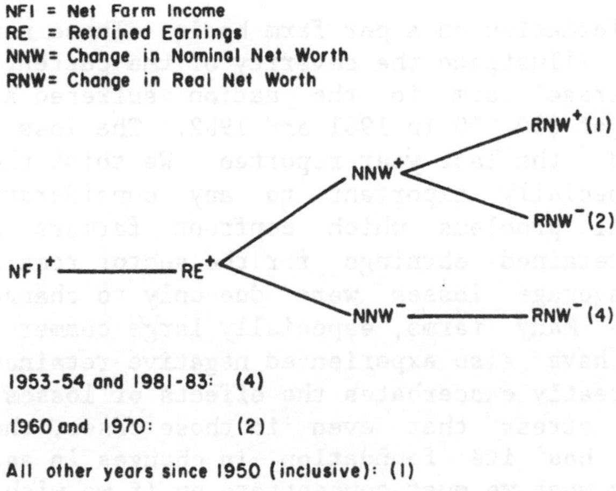

NFI = Net Farm Income
RE = Retained Earnings
NNW = Change in Nominal Net Worth
RNW = Change in Real Net Worth

$$NFI^+ \longrightarrow RE^+ \begin{cases} NNW^+ \begin{cases} RNW^+ (1) \\ RNW^- (2) \end{cases} \\ NNW^- \longrightarrow RNW^- (4) \end{cases}$$

1953-54 and 1981-83: (4)

1960 and 1970: (2)

All other years since 1950 (inclusive): (1)

Figure 4.5. Historical Sector Cash Flow Situations.

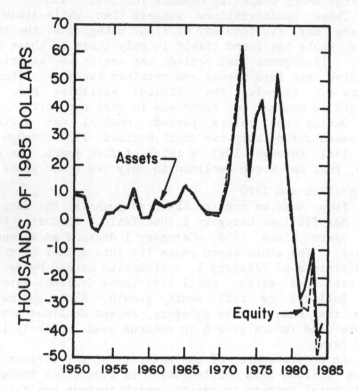

Figure 4.6.   Average of Farm Returns to Assets and Equity
              (Return to Assets Minus Interest)

this information on a per farm basis. These national statistics illustrate the severity of the current situation.
The average farm in the nation suffered a decline in
equity of $30,000 in 1981 and 1982. The loss was smaller
in 1983, the last year reported. We think these figures
are especially important to any consideration of the
financial problems which confront farmers at present.
Since retained earnings for the sector remain positive,
these average losses were due only to changes in asset
values.   Many farms, especially large commercial operations, have also experienced negative retained earnings.
That greatly exacerbates the effects of losses in equity.
But we stress that even in those cases, the essential
problem has its foundation in changes in asset values.
That is what we must concentrate on if we wish to develop
an effective and lasting solution to the financial problems of farmers.

Measuring Financial Stress on Different
Types and Sizes of Farms

The chief problem which analysts wishing to measure financial stress face is the matter of a convenient source of adequate information. Fortunately, there are two readily available sources which can provide the kind of material we need--the Illinois Farm Business Management Association (FBFM) and the Farm Credit Association AGRIFAX program. Both provide record keeping services for members, and both provide annual summaries of the information they gather. For our purposes, those summaries provide invaluable insight into the nature and degree of cash flow problems on farms of various kinds and of various sizes. Using each source in turn, we can derive a measure of retained earnings and capacity for debt repayment.

Capacity to repay debt is a less restrictive measure of well-being (vis-a-vis retained earnings), but is of extreme importance to agricultural lenders as well as "future" borrowers. Capacity for debt repayment is simply retained earnings plus the non-cash expense, depreciation. Even when retained earnings are negative, it may be possible to make scheduled principle payments or otherwise reduce debt through allowing gross investment to remain below the level of capital consumption or depreciation. Negative net investment in this manner can provide an orderly means for reducing assets on a farm (or in the sector) where conditions indicate that over-investment has occurred in the past. Many farmers, whose retained earnings are negative but capacity for debt repayment is positive, will find that this analysis covers their situation and suggests a useful financial strategy.

Illinois FBFM Members

We will begin with the Illinois farm records data. These farms tend to be larger than average farms with a moderately higher level of management skill than average in the nation. But, the Illinois farm record population has much less debt than the AGRIFAX population which is studied subsequently.

The Illinois Farm Business Management Association, along with the University of Illinois Cooperative Extension Service, performs a record keeping service for 7,977 Illinois farms. Its clients tend to be larger than the state average and exhibit somewhat more skillful

67

management than the national average. FBFM enrolls roughly 20 percent of the farms over 500 acres, these commercial farms averaging slightly more than twice the mean acreage of the state's farms. They account for the majority of the state's agricultural output. An important difference between FBFM and AGRIFAX farms--the FBFM clients carry much less debt.

Calculation of average retained earnings for all FBFM members in the last five years (see Table 4.6) shows the average FBFM member has positive net farm earnings. However, when added to non-farm earnings, these net farm earnings totaled less than the amount spent for family living expenses, and income taxes, in four of the past five years. That is, retained earnings have been negative except in 1983 when a small surplus was achieved. In real terms, net farm income has trended upward since 1979 although it was quite volatile from year to year. Real total income was more stable and had a strong upward trend. Consumption increased substantially (48 percent in real terms since 1979) with reduced taxes an important source of funds. Real retained earnings trended sharply upward from a negative $9,064 in 1979 to a positive $1,964 in 1983. It is interesting to note that both extremes in living expenditures still had negative retained earnings in 1983 while the middle group had larger positive retained earnings.

Strategies with regard to both debt and investment appear to be quite different for those groups. For the whole FBFM population, net investment has been rapidly curtailed since 1979. Net borrowings were curbed at an even faster rate. However, the problem which these statistics delineate may not be as severe as it first appears. In each case, funds have been available to reduce debt, and the trend is toward greater reductions. It was suggested above that this case might describe a population undergoing disinvestment.

In fact, net investment for the FBFM population has been declining generally since 1979 and was negative for the first time in 1983. Apparently there has been a substantial and on-going adjustment process throughout this period with operators changing owned asset levels and complements, as well as enterprise mixes. Some operators are saving and making positive net investments through reduced consumption expenditures (as in columns sorted by living expenses reveal). Others have maintained consumption and are disinvesting at a more rapid rate than average. By 1983, these processes reduced gross capital expenditures

Table 4.6. Retained Earnings for an Average Illinois Farm Business Records Member.[a]

(Current Dollars)

		1983[b]			1982[b]		1981	1980	1979
	All	High-Third	Low-Third	All	High-Third	Low-Third	All	All	All
Gross Farm Income	+	148671	183637	114779	149695	172591	128899	136447	140892	147324
Non-Factor Operating Exp.	−	84680	102511	68962	90769	101034	84054	80284	83684	87549
Interest	−	22812	36208	16018	22644	33447	17555	16619	14359	12497
Capital Consumption	−	15330	19705	14998	20720	32974	15933	22232	18155	31093
Net Farm Income	=	25841	25213	14801	15562	5136	11357	17312	24694	16185
Off-Farm Income	+	6873	10735	4632	8202	14171	7334	4766	3617	3804
Total Income	=	32714	35948	19433	23764	19307	18691	22078	28311	19989
Living Expenses	−	26495	37516	18124	25912	34733	16984	25912	23432	23486
Taxes	−	4255	3183	2469	4302	4782	3056	6008	6130	8414
Retained Earnings	=	1964	−4751	−1160	−6950	−20208	−1349	−9842	−1251	−11911
Net Borrowing	+	−3546	1914	8816	16549	31254	15552	18036	22422	40110
Net Investment	−	−1582	−2837	7656	9599	11046	14203	8194	21171	28199
	=	0	0	0	0	0	0	0	0	0
Retained Earnings	+	1964	−4751	−1160	−6950	−20208	−1349	−9842	−1251	−11911
Capital Consumption	+	15538	19705	14998	20720	32974	15933	22232	18115	31093
Capital for Debt Repayment	=	17302	14954	13838	13770	12766	14584	12390	16904	19182

Table 4.6 (continued).

	1983[b]			1982[b]			1981	1980	1979
	All	High-Third	Low-Third	All	High-Third	Low-Third	All	All	All

(1983 Dollars)

Item	1983 All	1983 High-Third	1983 Low-Third	1982 All	1982 High-Third	1982 Low-Third	1981 All	1980 All	1979 All
+ Gross Farm Income	148671	183637	114779	144006	166032	124001	124167	118208	112114
− Non-Factor Operating Exp.	84680	102511	68962	87320	97195	80860	73058	70211	66625
− Interest	22812	36208	16018	21784	32176	16888	15123	12047	9510
− Capital Consumption	15338	19705	14998	19933	31721	15328	20231	15232	23662
= Net Farm Income	25841	25213	14801	14971	4941	10925	15754	20718	12317
+ Off-Farm Income	6873	10735	4632	7890	13632	7055	4337	3035	2895
= Total Income	32714	35948	19433	22861	18573	17981	20091	23753	15212
− Living Expenses	26495	37516	18124	24927	33413	16339	23580	19659	17873
− Taxes	4255	3183	2469	4620	4600	2940	5467	5143	6403
= Retained Earnings	1964	-4751	-1160	-6684	-19440	-1296	-8956	-1048	-9064
+ Net Borrowing	-3546	1914	8816	15920	30066	14961	16143	18812	30524
− Net Investment	-1582	-2837	7656	9234	10626	13663	7456	17762	21459
=	0	0	0	0	0	0	0	0	0
+ Retained Earnings	1964	-4751	-1160	-6694	-19440	-1296	-8956	-1048	-9064
+ Capital Consumption	15538	19705	14998	19933	31721	15328	20231	15232	23662
= Capital for Debt Repayment	17302	14954	13838	13247	12281	14030	11275	14182	14598

a Derived from Illinois Farm Business Records, University of Illinois.
b Records were sorted into high- and low-third categories according to total living expenses.

70

to less than the amount by which existing capital assets were used up during the year. This process would continue until the farms are set on an entirely new course consistent with a new state of equilibrium in the farm economy.

We may safely assume that Illinois FBFM members are, in general feeling some stress. Category 6 in our ranking of financial stress levels (Table 4.2) seems to describe the average FBFM member. While he is not experiencing extreme stress, his situation is serious--on a level below the median which Category 5 represents. As with the sector as a whole, real net worth on these farms has declined during the past several years because of declining asset values. Adding additional stress for FBFM members is the series of years with negative retained earnings.

Cashflows by Type and Size of Illinois Farms

Although the 1983 FBFM summary report contains only a portion of the necessary data for constructing a detailed cashflow breakdown by type and size of farm, we can--by assuming some reasonable levels of net farm draw for living expenses[9], an average interest rate on debt and the level of debt as a proportion of assets--estimate the retained earnings and debt repayment capacities for different types and sizes of farms. The nine tables--4.7 through 4.16--present these estimates for farms in northern and central Illinois and for those in the southern part of the state.

Methodology

A detailed consideration of Table 4.7 will make clear how we generate the estimates of retained earnings and capacity for debt repayment in these tables. The first value, for each kind of farm, is net farm income before interest, NFIBI. This comes from the <u>1983 Summary of Illinois Farm Business Records</u> (SIFBR). The second value, which we use in deriving debt repayment capacity, is the amount of non-cash expense for capital consumption. Third, we show the sample average amount of assets which this kind of farm controls.

We assume an average interest rate of 10.5 percent. And we use that and assets controlled to generate interest

71

Table 4.7. Average Net Farm Income, Retained Earnings and Debt Repayment Capacity for Grain Farms (High Soil Rating, Northern and Central Illinois) by Size and Management Returns and Under Different Levels of Financial Leverage, 1983.

Size (Total Acres)	D/A	180-339	340-799	800-1199	>1199	All	340-799*	
Mgt. Return							Low 25%	High 25%
Number of Farms		111	545	116	49	821	136	136
NFIBI		42518	84455	154568	250292	98589	42821	131046
Dep		11994	20949	31819	55393	23329	23981	20186
Assets		945266	1761670	3085865	4743946	2016380	1713132	1921909
Interest	.2	19850	36995	64803	99623	42344	35976	40360
	.5	49626	92488	162008	249057	105860	89939	100900
	.7	69477	129483	226811	348680	148204	1256915	141260
NFI	.2	22667	47460	89765	150669	56245	6845	90686
	.5	-7108	-8031	-7438	1235	-7269	-47118	30146
	.7	-26959	-45026	-72243	-98388	-49613	-83094	-10214
Net Non-Farm Draw		10000	18000	25000	32000	20000	18000	18000
Retained Earnings	.2	12667	29460	64765	118669	36245	-11153	72686
	.5	-17108	-26031	-32438	-30765	-27269	-65118	12146
	.7	-36959	-63026	-97243	-130388	-69613	-10194	-28214
Cap for Debt Repay	.2	14661	48409	101584	186062	59574	10826	90872
	.5	-15114	-7082	4379	36628	-3940	-43137	30332
	.7	-34965	-44077	-60424	-62995	-46284	-79113	-10028

* These farms are sorted by estimated returns to management.

72

costs assuming 20%, 50%, and 70% levels of financial leverage. It is worth noting that in comparison of these costs with the average interest cost from Table 4.6, which was $22,812 in 1983, indicates that many farms have less than 20 percent debt, or pay less than 10.5 percent interest. Finally, we subtract the estimated interest costs in turn from NFIBI to obtain three estimates of net farm income--one for each degree of financial leverage.

Next we need to derive an estimate of retained earnings. However, to do so we need to estimate net non-farm draw from the farm business, off-farm earnings less family living expenditures and taxes. The 1983 summary did not include estimated net non-farm draws for each farm, although it did report an average 1983 cash living expenditure of $26,495 for a subsample of 257 central Illinois farm-operator families. And, for that subsample, it reported an average tax of $4,255 and a net non-farm taxable income of $6,873. However, national statistics (Figures 2.3 and 2.4) suggest that off-farm earnings decrease with farm size. Since these Illinois farms tend to be larger than average, those numbers may be on the low side. Further, preliminary reports by the Economic Indicators Branch, ERS/USDA, suggest that non-farm earnings differ across farm types according to the amount and timing of labor requirements. For example, non-farm earnings are less on dairy farms since the operator must tend to farm chores at regularly scheduled times during the day. Information on net draw from the farm business was available by size of operation for grain and dairy operations in the 1983 AGRIFAX summary report, so we also used that to guide our assumptions concerning this variable. Taken together, this information allows us to estimate net non-farm draw with confidence.

Subtracting the assumed net draws from net farm income gives estimates for retained earnings under each level of financial leverage. Finally, adding back the non-cash depreciation expense to retained earnings gives estimates for the amount of cash available for debt repayment.

These estimates, though based upon assumptions about levels of net draw for living expenses, serve well to illustrate the plight of highly leveraged individuals and to indicate the tradeoff between economies of scale, and the cost of debt. The ability to maintain cash flow when income declines varies significantly according to size, amount of debt, and variety of income sources. As these analyses suggest, expansion which entails the greater risk of substantial debt financing may be significantly less profitable than that which can be accomplished without acquiring additional debt.

Grain Farms

In general, northern and central Illinois grain farms exhibit some interesting earnings tendencies. Overall, these farms show strong net farm income results for 1983. As the 1983 Summary says:

> The net farm income (before interest) of grain farms having no livestock in northern and central Illinois (340 to 499 acres) averaged $61,865 in 1983 ... This income is $11,092 above that of 1982 and approaches the pre-1981 period. Even though corn yields averaged 55 bushels per acre below yields in 1982 and soybean yields 4 bushels below, crop returns were boosted $20 per acre above those of 1982 because of 8 to 10 percent higher selling prices for grain sold during the year; year-end inventory prices that were 40 percent higher; government acreage reserve and PIK program payments. The higher crop returns plus 5 percent lower cash-operating expenses accounted for the net farm income (before interest) increase (SIFBR, p. 7).

For a variety of reasons, the final picture becomes far more complex than consideration of only NFI makes it seem. In general, the interplay of amount of debt, size, and soil quality produces some curious results. Bigger farms generated more income, and low debt farms were more profitable--as we might expect. But farmers with poorer soil showed better ability to handle debt than those with better soil, which somewhat compensates their weaker income.

Low debt crop farms in northern and central Illinois made substantial income on average in 1983 (Table 4.7). The average net farm income for those grain farms with high soil ratings was $56,245 in 1983 (column 5 in Table 4.7) with debt equal to 20 percent of assets. Fifty percent debt would raise interest costs by $63,516 and leaves net farm income a negative $7,269. Financial leverage at the 70 percent level implies a very substantial loss. The largest class of farms though (those with sales averaging $500,600) had a small positive net farm income at the 50 percent debt level. The final two columns, which sort farms by level of estimated returns to management, show that some farms did well with higher debt levels while others did poorly with low debt levels.

Also not surprisingly, net farm income increases rapidly with farm size for farms with low debt. But, the relationship changes markedly as the level of debt and, so, interest cost increases. With 50 percent debt, net farm income tends to be unchanged as farm size increases. But with 70 percent debt, losses mount rapidly with increasing size.

Table 4.8 develops a comparable profile for grain farms with lower quality soil. Clearly, these farms earn less income than those with better soil. However, the lower ability to generate income principally reflects a difference in asset value. Poorer land is worth less and produces less, but interest costs are also less. An especially interesting result of the interplay of these factors is that these farms can stand a somewhat higher level of debt than comparable farms with high quality land. That is, a highly leveraged farmer with poor soil suffers less loss than a highly leveraged farmer with better soil.

Grain Farms in Southern Illinois: Southern Illinois farms of all kinds had a bad year in 1983 (Table 4.9). The effects of the drought were quite severe in that area, especially for grain farmers. Corn yields averaged only 65 bushels per acre. In nominal dollars, net farm incomes for farms in the 340 to 499 acre size category were the lowest since 1971 and 28 percent below 1982.

After subtracting net non-farm draw and adding back depreciation, we find that all farms with 30% debt, or less, can reduce debt with funds generated within the farm business. Only larger farms or those ranking in the top 25 percent by income, are capable of reducing debt if they have 50 percent leverage. No farms have positive retained earnings or funds available to repay debt when leverage exceeds 70 percent.

Hog Farms: Net farm income for northern and central Illinois hog farms reacts to size and debt much as it does for grain farms (Figure 4.10). Where debt is greater than 50 percent of assets, hog farmers have negative net farm income, retained earnings and capacity to repay debt. On low quality soil (Table 4.11), retained earnings are negative for all but the largest farms (those over 800 acres).

Hog farms in southern Illinois (Table 4.12) had lower returns, even in proportion to assets, than those in northern and central Illinois. The average southern

Table 4.8. Average Net Farm Income, Retained Earnings and Debt Repayment Capacity for Grain Farms (Low Soil Rating, Northern and Central Illinois) by Size and Management Returns and Under Different Levels of Financial Leverage, 1983.

Size (Total Acres)	D/A	180-339	340-799	800-1199	>1199	All	340-799	
							Low 25%	High 25%*
Mgt. Return								
Number of Farms		111	545	116	49	821	136	136
NFIBI		33760	71919	124859	230486	86789	29985	119371
Dep		13251	21653	36892	50718	25208	25008	19892
Assets		793105	1505188	24475853	4137191	1757972	1519034	1628185
Interest	.2	16655	31609	51993	86881	36917	31900	34192
	.5	41638	79022	129982	217202	92294	79749	85480
	.7	58293	110631	181975	304084	129211	111649	119672
NFI	.2	17105	40310	72866	143605	49872	-1913	85179
	.5	-7878	-7103	-5123	13283	-5504	-49764	33891
	.7	-24533	-38712	-57116	-73596	-42420	-81662	-300
Net Non-Farm Draw		10000	18000	25000	32000	20000	18000	18000
Retained Earnings	.2	7105	22310	47866	111605	29872	-19913	67179
	.5	-17878	-25103	-30123	-18716	-25504	-67764	15891
	.7	-34533	-56712	-82116	-105596	-62420	-99662	-18300
Cap for Debt Repay	.2	20356	43863	84758	162323	55080	5093	79071
	.5	-4627	-3450	6769	32001	-296	-42756	35783
	.7	-21282	-35059	-45224	-54878	-37212	-74654	1591

* These farms are sorted by estimated returns to management.

Table 4.9. Average Net Farm Income, Retained Earnings and Debt Repayment Capacity for Illinois Grain Farms by Size and Management Returns and Under Different Levels of Financial Leverage, 1983.

Size (Total Acres)	D/A	180-339	340-799	800-1199	>1199	All	340-799* Low 25%	340-799* High 25%
Mgt. Return								
Number of Farms		61	322	164	99	646	81	81
NFIBI		14506	27497	55893	117443	47264	-3720	62954
Dep		11404	21849	35264	64570	30816	29804	17347
Assets		541292	1008122	1725057	3091585	1465340	1173205	992565
Interest	.2	11367	21170	36226	64923	30772	24637	20844
	.5	28418	52926	90565	152308	76930	61593	52110
	.7	39785	74097	126792	227231	107702	85230	72954
NFI	.2	3139	6326	19667	52520	16492	-28357	42110
	.5	-13910	-25429	-34672	-44865	-29666	-65313	10844
	.7	-25277	-46509	-70897	-109788	-60438	-89950	-9998
Net Non-Farm Draw		10000	18000	25000	32000	20000	18000	18000
Retained Earnings	.2	-6861	-11672	-5333	20520	-3508	-46357	24110
	.5	-23910	-43429	-59672	-76865	-49666	-83313	-7154
	.7	-35277	-64598	-95897	-141788	-80438	-107950	-27998
Cap for Debt Repay	.2	4543	10175	29931	85090	27308	-16553	41457
	.5	-12506	-21580	-24408	-12295	-18850	-53509	10191
	.7	-23873	-42749	-50533	-77218	-49622	-78146	-10652

* These farms are sorted by returns to management.

Table 4.10. Average Net Farm Income, Retained Earnings and Debt Repayment Capacity for Hog Farms by Size and Months of Labor (High Soil Rating, Northern and Central Illinois) and Under Different Levels of Financial Leverage, 1983.

Size (Total Acres)	D/A	60-259	260-499	500-799	>799	All	By Months of Labor	
							21-27	31-39
Months of Labor								
Number of Farms		34	76	44	22	176	26	17
NFIBI		29625	58245	76097	159297	69811	66471	86692
Dep		15687	26754	41559	74485	34284	32122	47864
Assets		738446	1305645	2114496	3894486	1721889	1569409	2507625
Interest	.2	15507	27418	44404	81784	36160	32958	52660
	.5	38768	68546	111011	204460	90399	82394	131650
	.7	54276	95965	155415	286245	126559	115352	184310
NFI	.2	14118	30826	31692	77513	33651	33513	34032
	.5	-9143	-10301	-34914	-45162	-20588	-15921	-44958
	.7	-24649	-37718	-79318	-126946	-56746	-48880	-97618
Net Non-Farm Draw		18000	25000	28000	32000	25000	28000	32000
Retained Earnings	.2	-3882	5826	3692	45513	8651	5513	2032
	.5	-27143	-35301	-62914	-77162	-45588	-43921	-76958
	.7	-42640	-62718	-107318	-158946	-81746	-76880	-120618
Cap for Debt Repay	.2	11805	32580	45252	119998	42935	37635	49896
	.5	-11456	-8547	-21355	-2678	-11304	-11799	-29094
	.7	-26962	-35964	-65759	-84461	-47462	-44758	-81754

78

Table 4.11. Average Net Farm Income, Retained Earnings and Debt Repayment Capacity for Hog Farms by Size and Months of Labor (Low Soil Rating, Northern and Central Illinois) and Under Different Levels of Financial Leverage, 1983.

Size (Total Acres)	D/A	60-259	260-499	500-799	>799	All	By Months of Labor	
							21-27	31-39
Months of Labor								
Number of Farms		34	76	44	22	176	26	17
NFIBI		27512	35866	55217	93069	48131	55987	61679
Dep		18430	26338	38074	51105	32135	36091	44657
Assets		582715	1069728	1697935	2810056	1424269	1576969	1796508
Interest	.2	12237	22464	35657	59011	29910	33116	37727
	.5	30592	56161	89142	147528	74774	82791	94317
	.7	42830	78625	124798	206539	104684	115907	132043
NFI	.2	15275	13402	19560	34058	18221	22871	23952
	.5	-3080	-20293	-33924	-54457	-26643	-26802	-32636
	.7	-15316	-42759	-69581	-113470	-56551	-59920	-70364
Net Non-Farm Draw		18000	25000	28000	32000	25000	28000	32000
Retained Earnings	.2	-2725	-11598	-8438	2058	-6777	-5129	-8046
	.5	-21080	-45293	-61924	-86457	-51643	-54802	-64636
	.7	-33316	-67759	-97581	-145470	-81551	-87920	-102364
Cap for Debt Repay	.2	15705	14740	29634	53163	25356	30962	36609
	.5	-2650	-18955	-23850	-35352	-19508	-18711	-19979
	.7	-14886	-41421	-59507	-94365	-49416	-51829	-57707

Table 4.12. Average Net Farm Income, Retained Earnings and Debt Repayment Capacity for Southern Illinois Hog Farms by Size and Management Returns and Under Different Levels of Financial Leverage, 1983.

Size (Total Acres)	D/A	60-259	260-499	500-799	>799	All	By Months of Labor	
							21-27	31-39
Number of Farms		46	103	83	51	283	67	25
NFIBI		13625	17869	35208	58830	29646	27320	63113
Dep		16112	23554	33596	52207	30453	34157	51950
Assets		393371	759901	1183972	1963174	1041542	1148204	1425626
Interest	.2	8261	15958	24863	41227	21872	24112	29938
	.5	20652	39895	62158	103067	54681	60281	74845
	.7	28913	55853	87022	144293	76553	84393	104784
NFI	.2	5364	1911	10344	17603	7774	3208	33175
	.5	-7025	-22024	-26950	-44235	-25033	-32959	-11732
	.7	-15286	-37982	-51812	-85463	-46907	-57071	-41670
Net Non-Farm Draw		18000	25000	28000	32000	25000	28000	32000
Retained Earnings	.2	-12634	-23087	-17655	-14305	-17226	-24792	1175
	.5	-25025	-47024	-54950	-76235	-50033	-60959	-43732
	.7	-33286	-62982	-79812	-117463	-71907	-85071	-73670
Cap for Debt Repay	.2	3476	465	15940	37810	13227	8365	53125
	.5	-8913	-23470	-21354	-24028	-19580	-26802	8218
	.7	-17174	-39428	-46216	-652156	-41454	-50914	-21720

Illinois hog farm has assets equal to only 60% of the assets for a parallel farm in the northern part of the state, but its net farm income before interest amounts to only 42 percent. The effect of that is that relative to farms in the rest of the state, southern farms generate lower retained earnings and have less capacity to repay debt. However, highly variable soil quality in southern Illinois produces considerable variability among these results (SIFBR, p. 9).

Except in 1982, Illinois hog farms have had substantial negative returns to management in four of the past five years according to SIFBR calculations. Hog profits dropped because of overexpansion of hog numbers in 1983. When combined with low crop yields and continued high interest charges, management returns in 1983 were the second lowest on record (SIFBR, p. 8).

Dairy Farms: Costs have stabilized, milk prices have declined, so negative returns to management have increased each year since 1980 (SIFBR, p. 16). In spite of this, large herd groups did relatively well except in southern Illinois (Table 4.13). Still, those with little off-farm earnings or those who must borrow large amounts of money will not be able to stand long periods with conditions like those the dairy people experienced in 1983. In southern Illinois, estimates of retained earnings are negative for all sizes of farms, even those with low debt.

Beef Farms: In recent years, the picture for Illinois beef farmers has been even worse than that for dairymen. As the 1983 Summary says:

> For five consecutive years and for six of the last eight years, these farms have had negative returns for management exceeding $29,000 [calculated by charging opportunity costs for all assets and labor]. Historically, returns on beef farms have fluctuated between high and low profit years. Since 1974, there have been only two profit-making years--1975 and 1978. Higher management returns for 1983 resulted primarily from lower interest charges [due to decreased asset values]. Increased livestock sales, lower replacement cost for feeders, and 6 percent lower operating expenses offset the increased cost for depreciation [due to ACRS] and lower inventory values (SIFBR, p. 8).

For the most part, as a result of that situation, "Illinois farmers who maintain a beef-cow herd do so as a

Table 4.13. Average Net Farm Income, Retained Earnings and Debt Repayment Capacity for Northern and Southern Illinois Dairy Farms by Number of Cows in Herd and Under Different Levels of Financial Leverage, 1983.

Area of State	D/A	Northern Illinois				Southern Illinois			
Number of Cows		10-39	40-79	>79	All	10-39	40-79	>79	All
Number of Farms		46	103	83	51	46	103	83	51
NFIBI		23362	39060	59645	40070	9714	28111	48863	33224
Dep		17305	24213	46402	27370	16158	31124	43490	33740
Assets		747647	887172	1385219	961233	622318	844117	1295102	975387
Interest	.2	15700	18631	29090	20186	13069	17726	27197	20462
	.5	39251	46576	72724	50465	32672	44316	67993	51155
	.7	54952	65207	101814	70651	45740	62042	95190	71617
NFI	.2	7661	20429	30555	19884	-3353	10384	21666	12762
	.5	-15889	-7516	-13077	-10393	-22956	-16205	-19128	-17931
	.7	-31590	-26147	-42168	-30579	-36026	-33930	-46325	-38393
Net Non-Farm Draw		10000	15000	22000	15000	10000	15000	22000	15000
Retained Earnings	.2	-2338	5429	8555	4884	-13353	-4615	-334	-2238
	.5	-25889	-22516	-35077	-25393	-32956	-31205	-41128	-32931
	.7	-41590	-41147	-64168	-45579	-46026	-48930	-68325	-53393
Cap for Debt Repay	.2	14966	29642	54957	32254	2803	26508	43156	31502
	.5	-8584	1696	11323	1975	-16798	-81	2360	809
	.7	-24285	-16934	-17766	-18209	-29868	-17806	-24835	-19635

82

supplemental enterprise to market non-salable feeds and labor" (SIFBR, p. 18). However, large scale operations in northern and central Illinois had relatively high retained earnings where debt was low (Table 4.14). In no case were retained earnings positive where debt was 50 percent or greater, although the larger farms with 50 percent debt did have debt reduction potential through not replacing depreciated capital. Southern Illinois beef farms had negative cash flow with even 20 percent debt.

Poultry and Sheep: Poultry and sheep operations are not really a significant part of the agricultural economy of Illinois, and the 1983 Summary does not contain enough information to develop useful estimates of retained earnings and the capacity to repay debt for those kinds of farms. In spite of that, the report includes observations that suggest poultry and sheep operations reflect trends observed elsewhere. Poultry producers are seeking economies of size: "The flocks averaged 13,783 hens. Poultry in Illinois is rapidly being concentrated in fewer but larger and more industrialized operations" (SIFBR, p. 17). For many farmers, sheep, like cattle, have become little more than a useful sideline: "Most Illinois farmers who keep sheep do so as a supplemental enterprise in order to market non-salable feeds and labor." "The average return per $100 of feed fed in 1983 was $78 for native flocks." (SIFBR, p. 18).

Part-Time Farms: One alternative to attaining the scale necessary for size economies is part-time farming where the operator also participates in off-farm endeavors. Although the 1983 Summary does not directly discuss these farms, it contains data for generating net farm income figures for farms utilizing less than 10 months of operator labor. Net farm income for these farms follows much the same pattern as grain farms (Table 4.15). Like their larger neighbors, these farmers could not support debt equal to 50 percent or more of assets given an average interest rate of 10.5 percent.

As in the other tables, in this series we must assume a level of net draw from farm income to support family living expenses in order to generate retained earnings estimates and a measure of capacity to repay debt. Unfortunately, no consumption expenditure data for similar kinds of farms in terms of operator time requirements, or other factors which could identify part-time farmers, was available from SIFBR or from the AGRIFAX program for these smaller units. However, Figures 2.3 and 2.4 suggest that average off-farm earnings on the smallest classes of farms

Table 4.14. Average Net Farm Income, Retained Earnings and Debt Repayment Capacity for Beef Cattle Farms by Size and Months of Labor (Northern and Southern Illinois) and Under Different Levels of Financial Leverage, 1983.

| | D/A | Northern Illinois | | | | Southern Illinois | | |
| | | Size (Acres) | | | | By Months of Labor | | |
		180-339	340-799	>799	All	21-27	31-39	All
Number of Farms		28	90	48	166	326	14	50
NFIBI		28356	61166	154758	82695	107509	124060	16184
Dep		20953	30161	49183	37000	40937	56947	15919
Assets		858693	1532978	3233476	1910563	2079649	2673050	851956
Interest	.2	18032	32192	67903	40130	43673	56134	17891
	.5	45081	80481	169757	100325	109182	140335	44728
	.7	63114	112674	237660	140455	152854	196469	62619
NFI	.2	10323	28973	86855	42565	63836	67926	-1707
	.5	-16725	-13915	-14999	-176330	-1672	-16275	-28542
	.7	-34756	-51506	-82902	-57760	-45345	-72409	-46433
Net Non-Farm Draw		10000	15000	20000	15000	16000	20000	15000
Retained Earnings	.2	323	13973	66855	27565	47836	47926	-16707
	.5	-26725	-34315	-34999	-32630	-17672	-36275	-43542
	.7	-44756	-66506	-102902	-72760	-61345	-92409	-61433
Cap for Debt Repay	.2	21276	44134	126038	64565	88773	104873	-788
	.5	-5772	-4154	24194	4370	23264	20672	-27623
	.7	-23803	-36345	-43719	-35760	-20408	-35462	-45514

Table 4.15. Average Net Farm Income, Retained Earnings and Debt Repayment Capacity for Part-Time Farms (Used Less Than 10 Months Labor) by Size and Soil Rating in Northern and Southern Illinois and Under Different Levels of Financial Leverage, 1983.

Farm Type & Soil Rating	D/A	Northern Illinois					Southern Illinois		
		Grain Low		Grain High		Livestock	Grain		Livestock
Size (Acres)		< 260	> 260	< 260	> 260	All	< 260	> 260	All
Number of Farms		28	90	48	166	11	33	26	14
NFIBI		14733	35164	19580	68028	12772	1542	20382	-4353
Dep		7177	13208	7891	16167	8378	7519	13717	8491
Assets		444480	923320	608526	1255466	406934	286829	671489	262199
Interest	.2	9334	19390	12779	26365	8546	6023	14101	5506
	.5	23335	48474	31948	65912	21364	15058	35253	13765
	.7	32669	67864	44727	92277	29910	21082	49354	19272
NFI	.2	5399	15774	6801	41663	4226	-4481	6281	-9859
	.5	-8602	-13310	-12366	2116	-8592	-13516	-14871	-18118
	.7	-17936	-32700	-25145	-24247	-17136	-19538	-28972	-23623
Net Non-Farm Draw		0	0	0	0	0	0	0	0
Retained Earnings	.2	5399	15774	6801	41663	4226	-4481	6281	-9859
	.5	-8602	-13310	-12366	2116	-8592	-13516	-14871	-18118
	.7	-17936	-32700	-25145	-24247	-17136	-19538	-28972	-23623
Cap for Debt Repay	.2	12576	28982	14692	57830	12604	3038	19998	-1368
	.5	-1425	-102	-4475	18283	-214	-5996	-1154	-9627
	.7	-10759	-19492	-17254	-8080	-8758	-12019	-15255	-15132

suffice for adequate living standards. Accordingly, we assumed net draw to be zero.

Given that assumption about draw, retained earnings on these farms are positive with 20 percent debt. The largest grain farms, operating on a part-time basis, 387 acres tillable on average, have substantial retained earnings and capacity to repay debt with 20 percent debt. Furthermore, these larger farms appeared viable although not highly profitable with 50 percent debt. Livestock operations were less profitable and had negative income in southern Illinois.

Summarizing the Illinois Farm Financial Situation for 1983

As we put all these readings together, an interesting, if troubling, picture emerges. Farming is a losing proposition for an average Illinois farmer possessed of average management skill, working a moderate sized farm, and carrying debt approaching 50 percent of current asset value. Even before factoring in capital losses on the value of assets, that average farmer experienced substantial net farm operating losses and negative retained earnings. Worse, more highly leveraged Illinois farms typically could not repay debt even through the drastic strategy of disinvestment. In cases where the farmer took the option of not replacing the depreciated value of assets, he could continue in business only through substantial net borrowing against his declining asset values. As our earlier discussion of the effects of leverage makes clear, this strategy has the potential to redouble the problem in the future. In short, that statistical summary defines the latent explosiveness inherent in the cash flow problem these farmers are facing.

Illinois farm operators with average management skills, moderate-sized farms, and debt approaching 50 percent of the current value of assets experienced substantial net farm operating losses and negative retained earnings even before considering capital losses on the value of owned assets. Those more highly leveraged farms were, in general, not able to repay debt even through the rather drastic means of disinvestment where none of the depreciated value of assets is replaced. Such farms continued in business through substantial net borrowing against generally declining asset values. Clearly, this strategy has the potential to aggravate the problem in subsequent years. That is the cash flow problem today.

A breakdown by level of debt provides further insight into financial stress. In general, across all types and sizes of Illinois farms studied, debt levels over 50 percent of assets are associated with negative net farm income in addition to real net worth declines and negative retained earnings. This condition represents the highest category of stress in Table 4.2. Production Credit Association members of the AGRIFAX record-keeping program represent a group of these more highly leveraged operators experiencing this type of problem. They are studied in the next section.

Many of these severely stressed farmers, as we have suggested elsewhere, rank among the more able producers in our farm population. Most appear to be average or better farm businessmen. So we cannot just write these people off on the basis that they have brought their problems on themselves through their own inability to cope, as have some observers. All of that suggests that we must develop a more thorough understanding of their situation if we are to locate a policy approach which can produce genuine solutions to farm financial problems. Fortunately, for our purposes, the information which the AGRIFAX record keeping program has compiled deals with a group of more highly leveraged farmers who are experiencing just such problems as the Illinois data focuses on.

PCA AGRIFAX Members

The Production Credit Association's AGRIFAX program focuses on a group of farmers who not only figure to be experiencing severe financial stress but who also seem in many ways attractive and valuable members of the farming community. As the PCA points out, compared with farmers in general, "PCA members of the AGRIFAX program are much larger operators, much lower equity operations, faster growing operations, and younger operators than a random sample in the Louisville District would show. However, those members on the program represent a fairly random sample of this group of members with respect to management ability" (PCA 1981, p. 4). Our study of these records concentrates on farms located in the states of Ohio, Indiana, Kentucky, and Tennessee. It contrasts 1981 farm records of 1250 members with 1983 records for 1050. These

87

records include a good cross-section of the typical large commercial dairy, hog, beef-feeding, and grain farms which the Fourth District serves. Also these data compare favorably with the performance of most of the rest of the PCA loan portfolio of the district.

Some general characteristics for these farms are:

	1981	1983
Average Assets	$791,277	$732,642
Average Liabilities	388,161	338,365
Average Owner's Equity	403,066	394,277
Percent Equity	51	54
Average Owned Acreage	272	NA
Average Cropped Acreage	576	NA

An important question involves the extent to which the size and management skill levels are enough to carry the existent high debt load.

We have generated retained earnings for all farms in the program from AGRIFAX data for 1983 and displayed the results in Table 4.16. The data has been sorted according to net farm earnings in an attempt to differentiate between different levels of management skill. However, this approach has limitations in study of a single year and is more useful for illustrating the wide variations across farms due to all sorts of factors, including luck.

As expected, AGRIFAX members have more severe cash flow problems than the average farm in the nation. Retained earnings are negative for the average member in each of the two years studied. Most of the difference is due to debt. The debt-asset ratio for the nation was around 17 percent during the late 1970s; during the same period AGRIFAX members averaged around 50 percent. Further, off-farm income contributed much less to overall earnings for AGRIFAX members than for the average farm in the nation. Each of these factors is associated with the size and age of the farm operation. Larger operations require greater amounts of assets and tend to have a larger portion of debt. Young and growing operations have had less time to accumulate equity. Larger operations allow less available time for off-farm endeavor. But, whether due to management, debt level, or simply luck, only about one-third of the members would currently have insufficient cash available to allow the possibility of reducing debt and proceeding on an organized plan of disinvestment by not replacing depreciated capital.

Table 4.16. Retained Earnings for All Farms in the AGRIFAX Program.[a]

	1983			1981		
	Low 1/3	Mid 1/3	High 1/3	Low 1/3	Mid 1/3	High 1/3
Gross Farm Income	169450	138400	256300	202149	144256	188483
Non-Factor Operating Expenses	136200	91050	195600	203963	109847	114178
Interest	45200	25000	35000	58774	31088	26439
Capital Consumption	26600	18650	25750	30815	19029	17833
Net Farm Income	-38550	3700	51200	-91403	-15708	30033
Off-Farm Income	4850	4750	4100	6755	4647	4348
Total Income	-33700	8450	55300	-84648	-11061	34381
Living Expenses	17900	16350	26300	22153	17759	23592
Taxes	650	1000	1150			
Retained Earnings	-55250	-8900	27850	-106801	-28820	10789
Net Borrowing	NA	NA	NA	62463	26798	26163
Net Investment	NA	NA	NA	-44338	-55618	-15374
Capacity to Reduce Debt	-25650	9750	53600	-75986	-9791	2459

1982 data was not readily available to the author.

a Sorted on net farm income.

Four key indicators of cash flow problems are presented for crop (Table 4.17) and dairy (Table 4.18) farms,
which are classified by size, financial leverage, and net
farm income. The indicators are three-year averages
(1981-1983) for net farm income, net non-farm draw,
retained earnings, and capacity to reduce debt. The form
of these tables illustrates the individual impacts of management skill, farm size, and financial leverage on cash
flow conditions. In this case, where we have averaged the
data for each member over three years, the sort by net
farm income will tend to distinguish between management
skill levels while avoiding to some degree the random,
year to year, variation.

Taking into consideration the order of the sorting
process, the results for both crop and dairy farms point
out the overriding importance of maintaining a high net
farm income. The old adage "Get good, and then get big"
is clearly valuable advice. In general, profits increase
with size for successful managers and losses increase with
size for unsuccessful managers. The group with lowest net
farm income lose regardless of size or equity level. For
the middle group, higher levels of equity are necessary to
attain positive net farm incomes, and even higher equity
levels in combination with increased size are necessary to
maintain positive retained earnings.

The adage seems very likely to apply to the use of
financial leverage as well. The AGRIFAX analysis provided
cash flow information on the basis of farms, for several
types, grouped by returns to unpaid labor and management.
We constructed this sorting variable by using an opportunity cost for equity capital of 10 percent, which is nearly
equal to the average national interest rate on debt.
Accordingly, we have avoided confounding of debt-related
problems with business profitability to some extent.
Renters on crop and dairy farms with the highest returns
to labor and management indeed had the lowest equity
ratios and so higher levels of financial leverage (see
Tables 4.19 and 4.20). But, the opposite generally held
for owners and owner-renters. This reflects the marginal
rates of return on different types of assets. Financial
leverage can be beneficial when money is borrowed to
acquire assets which have higher rates of return.
Machinery and direct operating costs such as seed, chemicals, and fertilizer have generally had higher income
returns than land. Those individuals with higher returns
to unpaid labor and management from beef feeding also had
substantially lower equity (Table 4.21) indicating that a

Table 4.17. Average Cashflow Indicators for AGRIFAX Crop Farms (1981 through 1983)[a] by Size.

Net Farm Earnings	Sales (Thousand $)	Percent Debt					
		>66		33-66		<33	
Low 1/3	<125	-33000	-41500	-24100	-34200	-9000	-14800
		8500	-28100	10100	-20000	51000	-2000
	125-200	-53400	-69600	-28300	-45500	-13900	-29600
		16200	-41800	17200	-20300	15700	-11500
	>200	-97900	-117800	-48100	-68800	-9500	-33300
		19900	-84000	20700	-29700	23800	3800
Mid 1/3	<125	-2900	-13300	2000	-9000	11400	-700
		10400	-3800	11000	2500	12100	9200
	125-200	-3100	-18500	12800	-4900	26800	5500
		15400	1700	17700	15400	21300	26900
	>200	-13300	-41300	17000	-5200	45300	21100
		28000	-9700	22200	29300	24200	55100
High 1/3	<125	24700	12600	27700	11600	28999	17700
		12100	22200	16100	23600	11200	27500
	125-200	45000	23600	48300	30700	69800	45700
		21400	41000	17600	49600	24100	66300
	>200	54600	22500	85500	58200	122400	81700
		32100	50800	27300	90300	40700	114800

Key: | Net Farm Income   Retained Earnings |
      | |
      | Net Non-Farm Draw   Capacity to Repay Debt |

[a] (Sort: Number of Cows then Percent Debt then Net Farm Earnings)

Table 4.18. Average Cashflow Indicators for AGRIFAX Dairy Farms (1981 through 1983)[a] by Size.

Net Farm Earnings	Number of Cows	>66		33-66		<33	
					Percent Debt		
	<50	-13200	-21200	-7700	-15000	1500	-8300
		8000	-9200	7300	-6200	9800	1400
Low 1/3	51-80	-26900	-39300	-10100	-23500	3600	-7000
		12400	-21600	13400	-5700	10600	11200
	81-150	-41600	-60900	-22000	-24900	13700	-4500
		19300	-32200	15400	-11000	18200	23500
	<50	5200	-4600	7800	-1300	13800	2200
		9800	4400	9100	7800	11600	11700
Mid 1/3	51-80	4100	-8500	13200	-6200	25600	9300
		12600	6300	19400	10200	16300	26500
	81-150	-9900	-28100	11900	-8200	45100	19200
		18200	-4600	20100	18800	25900	45000
	<50	21000	11300	23000	11500	29200	14800
		9700	19300	11500	21500	14400	24900
High 1/3	51-80	17100	4100	36800	19700	59500	20500
		13000	17200	17100	36000	39000	35100
	81-150	25200	12500	46300	12500	79200	44500
		12700	38200	22800	48100	34700	63780

Key:

Net Farm Income	Retained Earnings
Net Non-Farm Draw	Capacity to Repay Debt

[a] (Sort: Number of Cows then Percent Debt then Net Farm Earnings)

92

Table 4.19. Average Cashflow Indicators for AGRIFAX Crop Farms for 1983 by Tenure.

	Return to Unpaid Labor & Management		
	Lower 33%	Middle 33%	Upper 33%
Owner-Renter			
Equity	59%	49%	57%
Net Farm Earnings	-24565	11031	58965
Net Non-Farm Income	4071	4929	6540
Operator Labor Draw and Income Tax	26055	19428	24764
Retained Earnings	-46549	-3468	40741
Capital Available for Debt Repay	-10006	16384	66117
Renter			
Equity	54.6%	39.2%	38.9%
Net Farm Earnings	-41200	6449	54716
Net Non-Farm Income	2970	4135	1428
Operator Labor Draw and Income Tax	13695	14990	17718
Retained Earnings	-51925	-4406	38426
Capital Available for Debt Repay	-24379	7679	61118

93

Table 4.20. Average Cashflow Indicators for AGRIFAX Dairy Farms for 1983 by Tenure.

| | Return to Unpaid Labor & Management | | |
	Lower 33%	Middle 33%	Upper 33%
Owner-Operator			
Equity	48%	57%	63%
Net Farm Earnings	-26167	16273	36159
Net Non-Farm Income	5902	4127	4241
Operator Draw and Income Tax	22487	22497	30805
Retained Earnings	-42752	-2097	9595
Capital Available for Debt Repay	-8269	22520	29996
Owner-Renter			
Equity	48%	50%	60%
Net Farm Earnings	-33463	4620	33894
Net Non-Farm Income	8238	4628	5886
Operator Draw and Income Tax	23387	19567	26446
Retained Earnings	-48612	-10319	13333
Capital Available for Debt Repay	-8552	14069	36524

| | Return to Unpaid Labor & Management | |
	Lower 50%	Upper 50%
Renter		
Equity	56%	40.7%
Net Farm Earnings	-2862	30399
Net Non-Farm Income	2814	-156
Operator Draw and Income Tax	18640	23079
Retained Earnings	-18688	7164
Capital Available for Debt Repay	-748	38595

Table 4.21. Average Cashflow Indicators for AGRIFAX Livestock Enterprises for 1983 by Tenure.

| | Return to Unpaid Labor & Management | | |
	Lower 33%	Middle 33%	Upper 33%
Swine Farrow to Finish			
Equity	46%	48%	51%
Net Farm Earnings	-45315	-12845	17336
Net Non-Farm Income	4745	2290	1005
Operator Draw and			
Income Tax	16954	11653	21263
Retained Earnings	-57524	-22208	-2922
Capital Available			
for Debt Repay	-31642	-4412	14815
Feeder Pig			
Equity	42%	57%	50%
Net Farm Earnings	-44583	1850	26912
Net Non-Farm Income	4688	4536	3480
Operator Draw and			
Income Tax	10500	17282	16966
Retained Earnings	-50395	-10896	13426
Capital Available			
for Debt Repay	-26792	3931	29103

| | Return to Unpaid Labor & Management | |
	Lower 50%	Upper 50%
Beef Feeding		
Equity	73%	41%
Net Farm Earnings	-15705	47508
Net Non-Farm Income	7187	1024
Operator Draw and Income		
Tax	15318	13148
Retained Earnings	-23836	35384
Capital Available for		
Debt Repay	-69	52265

number of operators were successfully utilizing financial
leverage in that enterprise. For swine operators (Table
4.21), financial leverage and returns to unpaid labor and
management had the opposite relationship.

Summary

Perhaps the most striking information from the
AGRIFAX data is the extreme variability across farms. It
is worth noting that averages fail to show how bad 1983
was for some farmers or how good it was for others.
Sorted on net farm income, the upper one-third had earnings nearly $90,000 higher than the lower one-third. The
upper group had 20 cents of each dollar of production left
after paying expenses but the lower group lacked 23 cents
per dollar of production to pay all expenses. Net non-
farm income did not significantly contribute to earnings
of AGRIFAX members. A large proportion of these members
had negative capacity to repay debt, lost substantial cash
equity in 1983 (ignoring additional capital losses), and
had debts past due during the year. For those, the future
prospect grows even more bleak. Most of these individuals
are in the most severe category of financial stress.

Conclusions

Financial stress in the farm sector as a whole is
currently more severe than it has been at any time since
the 1930s. Although the sector has sufficient equity to
survive, the real value of funds available for transfer of
farm assets is quite low. This fact is associated with
the dramatic fall in asset values.

Financial stress among the 8,000 farmers who are members of the Illinois Farm Business Record Keeping System
is more severe than that for the nation. These farms tend
to be larger and to have more debt. In addition to losses
in net worth due to declining asset values, these farmers
have, on average, faced negative retained earnings in four
of the last five years (1979-1983). Net farm income for
the group has been low but is still positive. Debt
financing shows up as the primary sign of financial stress
in our study of these farms. A substantial disinvestment
strategy appears to be underway in this group. These
observations appear equally true across all kinds of crop

and livestock farms. Increasing farm size has the fundamental effect of magnifying the existing situation.

Members of the Farm Credit System Record Keeping
Association have greater than average debt and, corre-
spondingly, greater financial stress than the group discussed above. In addition to declining net worths and
negative retained earnings, net farm incomes tend to be
negative among this group. Breakdowns within this group
illustrate the great variation across individuals. Some
of this variation is due to management ability. Some is
due to chance. A large proportion of these commercial
farmers are losing equity at increasing rates each year.

In sum, results of this study suggest that a large
number of farmers in the midwest are in bad financial
straits. Major adjustments in the capital structure are
underway. Farm income has been too low to validate recent
asset prices. Disinvestment and debt reduction strategies
are being rapidly pursued. Asset values are falling. All
those who hesitate to disinvest continue to lose equity.
Declining equity enhances the incentive to follow a disinvestment strategy. Asset values fall further. At some
point, values will have fallen to a level where current
returns make these assets profitable investments. Tremendous changes in the distribution of wealth, and so, in one
of the primary necessary ingredients for successful farm
business operation, will have transpired. Many farmers
who previously had the requisite equity to operate a certain size farm will no longer possess that. They will
continue on a smaller scale or leave the sector. Others
will take over their function. Adjustment will continue
as individual farmers strive for a farm size and organization consistent with their new equity situation.

NOTES

1. Discounting only cash flows, the present value of
a land investment can be shown to be equal to the present
value of annual net profit--defined as the current return
to land plus the gain in land value during the year less
interest charged on the land value at the beginning of the
year.

2. When agricultural sector data are used we will
follow Melichar and exclude inventory changes. Thus,
there may be some mismatching of income and expenses.

This error is felt to be smaller than that induced by including changes in the market value of inventories, which would add an element of unrealized gains or losses as market values of crops and livestock change.

3. Again, inventory changes will be included in net investment when using the currently available sector data.

4. Off-farm income is included in our accounting because it is a major source of additional cash revenue for many farm households. Provision of services in off-farm endeavor may be permanently viewed as an additional revenue generating product on some farms. For others, off-farm income provides additional income stability and revenue return for excess labor until such time as the farm attains sufficient size for viable full-time farm employment.

5. While a constant or decreasing general price level is conceptually possible our historical experience would not indicate that this is likely.

6. The change in real net worth can be computed by simply adjusting a time series of nominal net worth with a general price level index and taking the first difference of the series.

7. For other uses one might choose higher initial levels of some of these financial variables, such as the economic concept of normal returns to labor and capital.

8. However, on a per farm basis, the declines in real net worth in 1960 and 1970 were outweighed by the reduction in number of farms.

9. Net farm draw for living expenses is basically family living expenditures plus taxes net of off-farm earnings.

5

A Farm Problem

Locating the Problem

As we work towards an understanding of the financial problems which beset so many farmers, it becomes increasingly apparent that we cannot regard the present crisis, by itself, as the problem. As we have urged with the phrase land boom-bust and throughout our discussion so far, the kind of understanding which can lead to genuinely effective policy can only derive from a consideration of an economic process which typically unfolds over a period of several decades. Further, we think that if any one section of that long process bears more responsibility for the instability which is now relatively more obvious, it is the years of steady growth leading to the boom. During those years, farmers prospered and the prosperity lulled many into concluding that this was a normal and unchangeable state of affairs. People notice the problem during the bust--when the bottom drops out for large numbers of farmers. Paradoxically, since the real problem has to do with the structure of the farm sector, the problem is always there--even when farmers seem to be doing the best financially. So long periods of growth and prosperity tend to erode their sense of caution, and many farmers overlook the strategies that could protect them from serious downturns. To see why that is the case, we need to begin with a consideration of the general mechanism of a land boom-bust, and how these events have operated in other times. We need to think about who gains from boom-busts and who loses. Then, we must understand the nature of the land market and the traditional policy roles.

A story of financial distress in the farm sector, like the one which the 1981-83 statistics outline, comes about whenever beliefs about the level of future returns to farm assets change and expectations of substantially lower returns lurk behind every market report. The general pattern includes a period of generally rising asset values, climbing general prices, and steadily rising commodity prices. The positive nature of this long initial period creates the belief that the sector will continue to grow and experience profitable operations. Second, there is a boom period when prices and asset values climb at a dizzying rate--a time, as many say while an ebullient mood prevails, when anyone can make money farming. Finally, a precipitous drop in asset values and prices marks the bust --the end of the sequence.

These patterns are frighteningly disruptive, since they cause chaotic wealth transfers with their accompanying allocative inefficiencies and personal pain. The problem, as we have pointed out, is that the structural organization of the farm sector, particularly its systems of ownership and financing, does not readily accommodate downward changes in beliefs about future farm asset returns of the magnitude which occur from time to time.

Because of that adjustment problem, many farmers suffer from overinvestment in land and depreciable assets --assets for which the farmers have paid too high prices and used too much debt financing. As a result, farmers have experienced huge capital losses. Many farmers have not been prepared to deal with such large losses. For many, who had gambled on magnifying potential profits through aggressive use of financial leverage, the outcome, instead, has been magnified losses. Those painful losses are what most people refer to when they speak of the farm problem.

Clearly, many farmers could be in better financial condition if the structure of the farm economy were such that it could react more flexibly to situations that develop in the land market. We do not pretend that wrong decisions can be avoided. Farmers who manage badly, or otherwise make wrong decisions consistently, will suffer whatever the structure of the farm economy. Rather, we would like to see an institutional development which recognizes that expectations for future income returns to assets are subject to change--that the good managers suffer along with the less good, not necessarily through any fault of their own, but because they lack the tools for responding to this situation which the structure of the farm economy causes. Presumably, given such tools, better managers could create a defense against the disruptive downturn and continue from a position of financial strength.

Booms and Busts--Past and Present

Invoking historical context ought not be necessary, on this as on other issues. Yet farmers, like other people, have a short historical attention span. Perhaps our accounting and strategic planning habits help constrain our horizons. Seldom, it seems, do we keep track of events more than a decade remote from the present moment. More often, five years defines the historical attention span.

Any survey of the literature uncovers reminders that land booms are not unique to our own time and that the havoc they wreak now has its parallel in earlier times. Warren and Pearson, writing about the period of the Great Depression, have observed the effects of such a pattern following the War of 1812 and the Napoleonic Wars:

>   As a result of financial inflation, the average value of land in fourteen counties in Pennsylvania rose from $53 per acre in 1809 to $111 after the War of 1812. Deflation reduced the prices to $38 in 1819. In the words of Gouge [on land prices following the Napoleonic War period]: "Farms rose in price from fifty to a hundred percent, and sank again as rapidly as they had risen. Thousands were reduced to poverty and a few rose to wealth on the ruin of their neighbors." (Warren and Pearson 1935, p. 382).

Time and location make little difference. The story from one era seems an echo from an earlier. Ohio farmland was $160 per acre in 1920, $94 per acre in 1929 and $59 per acre in 1933. If that land were mortgaged at 50 percent of value in 1920, then the farmer was insolvent in 1933 (Falconer 1934). Indiana farmland was $3,000 per acre in 1979 and $2,000 per acre in 1984. The insolvency rate in Indiana is high and growing. By 1925, 75 percent of the value of mortgage debt in Germany had been wiped out through the hyperinflation following World War I. For some time after the stabilization of that inflation, interest rates in Germany were 20-30 percent. High real interest rates following that period caused severe stress for German farmers (Lester 1934). By analogy, the value of mortgage debt in the U.S. was severely eroded during

the 1970s and again high real interest rates (12.5 percent
prime commercial lending rates with 4 percent inflation in
early 1985) are causing stress. The lesson is the same
each time, but for some reason, each generation must learn
it for itself (Chambers 1924). Melichar (January 1984)
notes that the booms have been infrequent, two in this
century and two in the last; but the effects of each
extended over several decades. Each boom resulted in substantial changes in the financial conditions of many
people. Many have benefited. But a significant number of
farmers have been in such dire financial straights after
each boom, specifically those highly leveraged in over-
valued land, that they have endured prolonged financial
stress or entered bankruptcy.

A survey of conventional opinion on land booms and
busts through the years suggests the conceptual orienta-
tion on which most policy considerations have been based.
Land booms have their roots in unusual demand often stimu-
lated by war and inflation. They usually build slowly,
the culmination of a long upward trend in farm income. A
sustained period of confidence and high earnings seems to
be necessary. It is also important that farmers have
accumulated earnings. Returns to land rise. Land prices,
the capitalized value of total returns, also rise.
Returns rise again and people entertain expectations for
further increases. As returns continue to increase, the
expectations for increasing returns are capitalized, ten-
tatively at first and then with greater confidence. Land
booms at this stage tend to be financed largely by previous owners through second mortgages. There are many win-
ners, few losers and much excitement on the way up.

But then the tide turns. Those caught at the end
suffer great losses. We often think that, like wars, land
booms are undesirable and unnecessary.

On Asset Values

We think it especially valuable to consider historical parallels to the present situation, for that gives us
a sense of how the pattern develops and what drives it. A
land boom-bust sequence, such a consideration shows, is an
extreme example of the price chaos which Warren and
Pearson discussed in their chapter "Price Chaos Caused by
Inflation and Deflation" in Gold and Prices, 1935. Their
argument is that because prices guide all production and

consumption, price chaos creates severe distortions throughout most aspects of life. The more developed the exchange system and the more specialized the economy, the more severe the distortions.

Recently, Irving Friedman has agreed with that earlier analysis, calling the chaos of inflation a worldwide disaster because "everything had to adjust to the new monetary conditions--including the church's attitude toward what was a 'just' price and what were 'usurious' interest rates" (1980, p. 4). In the face of the vacuum which this chaos brings about, Warren and Pearson point out, government tries to provide a substitute--not always with happy results:

> Since the price guide for consumption is lost, [government] committees have to regulate almost every human act, even to deciding whether eggs may be served on toast or whether the two may be served separately, as was ordered by the United States Food Administration" (Warren and Pearson 1935, p. 4).

During such a period, furthermore, primary (wholesale) prices fluctuate more rapidly than secondary (retail) prices. Farmers, as primary producers, are among those who experience the most extreme effects of the situation. In addition, Warren and Pearson claim that the length of time required to produce a product conditions the effects of price chaos. In their words: "The longer the period from beginning to the completion of an article, the more serious are the effects of deflation" (1935, p. 312). It takes an especially long time to produce a farm product. As a result, farmers are intensely interested in the condition of the monetary system, for their ability to operate profitably depends on the value of money. As Warren and Pearson say:

> It is no accident that farmers are more interested in the money question than are city persons. It is not because farmers are radical; they are, in fact, less radical than most other groups. It is that deflation places on them an intolerable burden. The farmers of England were so adversely affected by the deflation following the Napoleonic Wars that many inquiries were made into the state of agriculture. Again in the nineties, the unfortunate condition of English agriculture received the spiritual benefits of an

investigation.   Its physical state, however, was not improved until the value of money started to decline" (1935, p. 313).

Davis provides a similar analysis; and in his outline of the plight of agriculture through a boom-bust pattern, he calls attention to the way the apparently good times help to bring about the bad:

> The period of rising prices brought the farmers exceptional prosperity for the time. Rising costs lagged behind rising prices, and large profits were made.   Land values rose, and mortgage debts previously incurred became less burdensome. Many farmers sold out and retired. But, the farmers' prosperity--real enough as it was, though less in reality than in appearance--carried the seed of depression. It overstimulated farm production, and in agriculture especially it is notoriously easier to expand than to contract. The increased credit margin tempted many to buy farms at high prices or tempted established farmers to buy more land, and bankers did not properly discourage these tendencies. It encouraged the breaking up of new lands, the planting of orchards and vineyards, the building up of livestock herds and flocks. In part, these were the responses to patriotic appeals for greater production; in larger measure, I believe, they were the response to high prospective returns, largely unchecked by expectations of subsequent reaction.
>   After the war, the decline in the general level of prices was a severe blow to farmers. Prices declined faster than costs. Farm values declined, but mortgages did not (Davis 1939, p. 80-81).

A study of the earlier panics of 1837, 1873, and 1929 shows that a chaotic period in real estate is even worse than chaos in other aspects of the economy because when people carry heavy real estate debt, the panic takes a long time to work its way through the system--partly because people are slow to acknowledge the actual situation and also because there are few options for remedy:

> A stock market crash can be liquidated quickly, but the process of mortgage liquidation is very slow. For several years, borrowers and lenders both hold

on, expecting that prosperity is just around the cor-
ner. There are no market quotations on homes so that
neither party realizes the situation. When foreclosures do occur, properties are commonly taken over by
the creditors, partly because of the dearth of buyers
and partly because the creditors do not realize the
situation. Before it is cleared up, the creditors
must be willing to sell at market prices and the
properties must pass back into the hands of persons
who want them. Many creditors still held farms in
1932 that were taken over in 1921. These were held
because of the unwillingness to accept the loss,
expectations of profit or fear of breaking the market
(Warren and Pearson 1935, p. 242).

Land Boom-Bust Policy

Warren referred to 1928 as the eighth year of the
most serious agricultural depression ever known in the
U.S., which was caused, he thought, by financial deflation. Farmers, and industries depending upon farm custom-
ers, brought tremendous pressure to bear on government to
raise prices after the decline (Englund 1923). Wallace
and others advocated reinflation as a remedy. The inflation group argued that the Federal Reserve should cause
prices to again equal those at which farmers had incurred
the bulk of their debt so that "... debtors and creditors
would have a fair chance ... in the interest of justice"
(Dowrie 1938). With their suggestion that a doubling of
price levels would make farming a prosperous existence,
Warren and Pearson lent support to that group (Lester
1934). The government adopted relief measures, such as
the Reconstruction Finance Corporation, to maintain the
values of capital by sustaining interest payments
(Peterson 1933). Emergency credit (Wall 1933), credit
reorganization (Baird 1933), debt moratorium (numerous
articles in J.F.E., 1933), and federal land purchase programs (Grey 1938) represent some of the many proposals
designed to deal with tremendous financial losses brought
about by declining farm prices.

"The fool and his money are soon parted; but the public divides broadly into two classes, those who think
the process should be accelerated and those who think
it should be retarded" (Nourse 1925, p. 1).

The formulators of all of these policies intended to help farmers weather the crisis and to move the economy back toward a healthy condition. Unfortunately, all of these policies had the additional effect of retarding recognition of the real losses which resulted from the boom-bust cycle, of impeding even further effective reaction to the very real problems. The importance of our historical hindsight becomes apparent when we note that the proposals for dealing with the present crises closely resemble those from the 1930s.

Yet the question to intervene or not is far from simple. To see why, we need only consider the status of asset values as an important farm problem. Some might argue that this becomes an issue only for less than skillful managers. They would probably claim that the debt-free farmer who has no intention of selling his assets and who has based no other investment decisions on the contemplation of such a sale can view his huge current paper loss with a certain stoicism, especially so if he also took part in the huge paper gains during the seventies. However, even if a sale was never contemplated, declining asset values are a severe problem for farmers borrowing substantial amounts of capital. The paper loss can suddenly become a crucial factor to a lender as he sees the market value of his security erode. So asset values affect the ability to borrow in any case. The current problem is acute for those already leveraged when the downturn began, who could not or would not unleverage, and who could not show a positive cash flow in terms of retained earnings and capacity to repay debt. But asset value declines, even if just paper losses, pose a genuine problem throughout the farm sector.

Jonathan Harsh, writing tongue-in-cheek in a farm magazine, points out a current policy dilemma:

> We've got a string of proposed legislation, designed to rescue over-leveraged farmers. Only problem holding up enactment: how to draft measures to make sure Federal money gets to producers in trouble due to economic conditions, not poor management ("Harsh on Washington," Farm Futures, July 1984, p. 6).

Clearly, there are bad managers, and they may deserve to fail. Just as clearly, though, many perfectly good managers are in serious trouble. In another article Harsh quotes Neil Harl, professor of agricultural economics at Iowa State University, on the same issue:

> Such intervention is needed, [Neil Harl] insisted, because: 'There is no doubt in my mind but that the people who are exiting [from the farm sector] are well beyond the group that should exit because of management abilities, unless you take the position that borrowing money is bad management (Farm Futures, July 1984, p. 22).

Borrowing money is a basic aspect of financial management, and good financial management is paramount to good farm business management. It is far from obvious, however, that we know how to evaluate management decisions. We tend, naturally enough, to evaluate the decisions managers make on the basis of outcomes. Depending on the time when we decide to pass judgement, however, those evaluations may vary. For example, if a farmer had borrowed heavily to buy land in 1972, we would probably have said that was a poor decision. Certainly it looks wrong on paper, even now. Yet that farmer is most likely in good shape now. In contrast, a farmer who borrowed to buy land in 1979 or 1980 may have been making a move we would approve of on paper. Yet he may be experiencing great stress. Timing plays an important role in all of this, and it may be that neither decision was correct no matter what the results suggest. Along with other factors, Warren and Pearson suggest that date of birth is important.

> The financial success of an individual is due in part to his own ability and in part to the time when he was born. The young man who started farming after the Civil War and who showed the most energy and courage usually suffered the most. Prices declined so rapidly that it was difficult to maintain the capital, to say nothing of making a profit. From 1897 to 1913, prices of farm products rose rapidly. The young man who started farming early in this period and who wasn't heavily in debt was successful. The uninformed point with pride to the initiative and success of this individual, as compared with his father, when in reality the only extra credit that is due to him is the choice of the time to be born (1935, pp. 429-30).

It is not difficult to construct an argument that society is responsible for the current financial condition of the thousands of over-leveraged PCA members whose situations we studied in Chapter Four of this book. Some may argue that by maintaining, through national monetary and fiscal policy, a relatively stable and growing farm income throughout the 1950s and 1960s and then by increasing the rate of growth through higher and then higher inflation during the seventies, society sent strong signals to alert financial managers to invest. By abruptly changing the monetary part of that policy, by suddenly stopping the inflationary growth in returns, society invalidated the actions of those who heeded thirty years of signals. To be sure, simple reference to other boom-bust periods can counter that argument. But most farmers are too young to remember the last bust. So too, apparently, were a majority of economists, bankers, and PCA lenders. So one could argue that many of those in trouble today were simply born at the wrong time.

Ultimately, this brings us back to the notion that the basic financial problem which plagues our farm economy has a structural basis. We cannot blame it on current events, though foreign policy decisions can certainly be vexing. And we cannot blame farmers and claim they cause their own distress. No doubt some do. But many others are really victims of traditional policies and of the accident of time of birth. For many farmers currently in distress, the same strategies undertaken five or ten years earlier would have produced dramatically different results. However, if we do make an effort to understand the structural nature of this problem, it seems likely that we can devise strategies and policies which can help us cope effectively.

Wealth Transfer or Deadweight Loss?

As we think about this troublesome economic phenomenon, we must confront the question of what actually happens when the value of agricultural land increases and then declines. What we want to know is whether such a process creates wealth and then destroys it or merely redistributes wealth. The answer to that last question is, we think, "it depends."

In general, excluding allocative inefficiencies associated with any adjustment process, and assuming that the value changes are within the range of something called economic rent[1], we think changes in land value to be transfers. Consideration of some hypothetical cases makes clear our reasons for that.

Perhaps the event which alters expectations involves a technological breakthrough in production which would increase the future supply. If that happened, the future value of food would be lower than previously anticipated. As a result, farmland would command less economic rent, its value would fall, and farmers would experience a reduction of wealth.

However, the wealth of society as a whole would increase. Since society could buy more food for the same price or the same amount of food for less, society as a whole would be better off and would, in effect, experience an increase in wealth. To the extent that farmers are consumers of food, they would share in the benefits of lower priced food. But the loss of wealth from falling land values would no doubt exceed their gains in terms of food purchasing. Farm owners are net losers. Non-landowners gain from the technological breakthrough and from the transfer of wealth from landowners.

A second kind of event might at first seem less beneficial to society. If forecasters suddenly had reason to expect a severe and protracted economic slowdown, this would alter future demand for farm products. As in the other case, the future value of food would go down, and farm owners would suffer a loss. But in this case, the same amount of food would have a lower value to society. Therefore we surely would not claim society to be better off as a result of the change.

Yet there is a transfer of wealth here. Consumers benefit by the value of the decrease in economic rent to the land. In fact, we conclude that in both cases, economic events neither create nor destroy wealth. Rather, the changing economic rents for land merely transfer wealth among people in society.

Concentration of Effect on Landowners

The simple transfer of wealth due to changing land values obscures one aspect of the situation which is of paramount importance. During a wealth transfer, the effect concentrates on farmers. Changes in beliefs about future prices for primary agricultural products which cause wealth losses of, say $100,000, for an average U.S. farmer have only a very small effect for the average consumer in the U.S. and a trivial effect when spread over all the consumers of the world. Since large losses will have proportionately larger impact on farm welfare, it may well be that what appears to be an economically stable wealth transfer, given the entire economy, will have far different implications for the welfare of the farm economy--on which the effect concentrates. Awareness of that provides a rationale for policies of governmental, or other public, intervention.

Even within the farm sector, there is a concentration and magnification of the effect of declining asset values. As we have said, beginning and expanding farmers normally have substantially greater debt than average farmers, so they suffer far more than average when there are revisions of forecasts. This is true because the degree of financial leverage amplifies the effects of lower than expected returns. These people may have to undergo substantially greater adjustments in response to lower returns than older, more established farmers. They may even have to liquidate their entire business.

Clearly, that farmer loses, in this situation, more than the average consumer gains. It is especially troubling to many economists to note that in many cases the biggest losers are some of the most able young farmers.

Agricultural Adjustment Mechanisms: The Case of Farm Real Estate

This discussion allows a further refinement of what the term "the farm problem" should mean in this kind of policy discussion. As is common knowledge, the farm economy fluctuates between periods of rising land values and periods during which there are large downturns in those values. During the periods of increasing land values all farmers appear to prosper as they experience large real capital gains. When the downturn comes, though, the consequences are problematical. Some continue to do well, but many others find their family life disrupted, their farming operations bankrupted, and many rural banks fail. It is the uneven concentration of effects, the uneven ability of various segments of the farm sector to adjust to large swings in asset values, that forms the core of the farm problem that we are addressing.

These are not phenomena of recent origin. Our farm economy has suffered from major changes in asset values on several occasions. What is curious is that we have not done more to develop institutional structures within our economy to facilitate effective adjustment to these events. In part, that lack of development may result from the long duration of periods of stability and prosperity. People in all walks of life have little sense of history. As generations pass, people may well forget the significance of their fathers' or grandfathers' experience. The boom periods further obscure memory. As an athlete recently remarked, "when you're winning, nothing else seems to matter". Finally, government policy, with its focus on relatively superficial phenomena like commodity prices, may have impeded development of those institutions.

However that may be, as we focus on the matter of fluctuating asset values and farmers' abilities to adjust, especially to the downside swings, we gather a variety of insights. Chief among these is the realization that useful adjustment mechanisms must focus not on commodity prices (or on matters of income transfer and market stability), but on fluctuating asset values. The distinction is important. Schultz (1953) outlined two functions of commodity prices--prices determine the level of compensation for factors of production and they guide allocation of resources--and pointed out inherent difficulties in using price intervention· to change the outcome from one function without causing undesired outcomes from the other. Our goal is to allow both commodity prices and asset values to fluctuate according to market conditions while better adapting the system of economic organization to accomodate the fluctuations. The policy setting we envision combines ownership patterns, market organization, mortgagability, and price and income policy for farmland in a way which will result in far less disruption with each downturn in the land market.

Of course, any such policy must follow from the nature of the marketplace and the system of ownership for farm land.

Liquidity of Farmland

The concept of liquidity relates to the how easily we can sell an asset or convert it into cash. Typically, we measure liquidity in terms of the length of time of the

111

conversion process and the transaction costs. Economists can have trouble with the discussion of liquidity. Simply enough, when an asset is hard to sell, we can say it is illiquid. Yet, if the price were low enough, the market would clear, and the asset would be liquid. So an important issue in discussions of liquidity might be the question of what discount is necessary to allow the asset to sell quickly. It seems reasonable to view this discount as a transaction cost which varies with the time and effort, or cost, it takes to make the sale.

In the real estate market, though, even such a discount may not provide liquidity. In this market, information problems concerning what is available and how it compares with other land parcels and high transaction costs relative to other investment assets lead the factors which create a situation where a market clearing price simply is not achieved on a daily or weekly basis. We are forced to conclude that the land market lacks liquidity. This asset is hard to sell.

Moreover, the land market entails liquidity risk. Basically, that involves a situation in which low asset values correlate strongly with the circumstances which create the need to sell. That is, when a farmer most needs to sell, agricultural conditions are most likely to be such that land values are low. Liquidity risk differs from the situation where someone can sell an asset quickly by offering an additional discount. Liquidity risk operates at a far more basic level.

Heterogeneous Product

Land is not a homogeneous product and cannot be transported, so the market does not easily accomodate arbitrage over space. This implies that the real estate market performs as a collection of smaller localized markets, loosely connected by the (imperfect) substitutability of the farmland parcels in each area. While there may be a large number of buyers and sellers throughout the nation, each small segment may have few or no active sellers and few or no active buyers.

Transaction Costs

The real estate market requires substantial transaction costs. Information in this market is very costly.

112

The cost of employing a realtor to offer a property for sale provides a measure of the cost and shows how expensive it can be. The seller must communicate specific details about the property to as large a number of potential buyers as possible and must negotiate price and terms with each. Spatially distant buyers must spend time or employ agents to view the heterogeneous product. On both sides, the price of discovery is high relative to other markets.

Time delays often add to buyer or seller costs. The time between acceptance of an offer and closing is often six to eight weeks. Financing, legal, and other costs associated with a real estate transfer can also be significant. Often the seller cannot transfer items of value, such as a low interest rate on a "due on sale" loan, to the buyer.

Arbitrage Opportunities

Arbitrage opportunities are limited, risky, and costly. For the most part, the substantial transfer costs limit arbitrage to that over time. Since resolution of the arbitrage profit is in the fairly distant future it carries substantial risk. Transaction costs require that there be a fairly large gain for arbitrage to break even. Management and other costs associated with holding land can be significant for non-operator landlords, adding to the margin required for non-farmers to provide the arbitrage function. A large amount of capital is usually required and is tied up in an illiquid asset to arbitrage in real estate. Finally, some states have legislation preventing ownership of farmland by certain classes of individuals (particularly people who are not local farm operators) which limits the arbitrage function.

Liquidity Risk

In addition to the land market being a thin or illiquid market, land has substantial liquidity risk for farmers. Though it may have less liquidity risk for non-farm investors whose portfolios are more broadly diversified. This aspect of liquidity risk derives a notion of liquidity value. Generally, the liquidity value of an asset is the market value less transaction costs. Market values are uncertain, so liquidity values are uncertain.

This is the basis for liquidity risk. A farmer cannot know how much cash he can raise by selling an asset in the future because the price of the asset is uncertain.

However, the uncertainty itself is not the only aspect of liquidity risk. One must consider why the liquidity values of assets in a continuing business are important. Many of these assets will never be sold or will only be sold after their productive life has expired, but their liquidity value helps regulate cash flow. Most businessmen are acutely concerned about cash flow, for good reason. If all were to work as expected, cash flow would not be a problem for most firms. That is, if values of random variables met expectations there would seldom be more than seasonal cash flow difficulties. Cash inflows should on average easily handle cash outflows.

Unfortunately, we can not average the cash flow of good and bad times. Cash flow must be balanced in all years, good and bad. In extremely bad times, the liquidity values of some assets may be called upon to provide cash. The special problem related to liquidity risk is that some assets have a substantially greater probability of low market value, when the owner needs to sell, than the unconditional probability of low value. Land has very high liquidity risk for farmers. When a farmer needs to sell land there is a good chance that agricultural conditions are poor and that land values are low.

Land Market Dynamics

Featherstone and Baker, in studying the history of aggregate agricultural sector real estate in the U.S., have developed a model of real asset values which allows examination of asset value dynamics.[2] Attention to two categories of information generates insight concerning the dynamics of the farm real estate market: causal relationships and impulse response. We will consider each of these in this section. But before continuing, we need to provide some interpretive guidance.

In the long run, we firmly believe, returns to land determine the value of land. But here, our concern is with observed market prices for land. While returns to land exercise an ultimate control, in the short run, beliefs about the future stream of returns to land constitute the operative factor. It should be understood that nothing in this section is inconsistent with economic

114

theory regarding land valuation. However, economic theory has little to say about the beliefs which people in the real estate market hold about future returns to land. People's beliefs derive from subjective ideas about the future supply and demand of farm outputs. But, events in the future are unknowable. Beliefs about these events represent only people's guesses, and we know little about how people formulate these guesses. Nevertheless, they seem to correlate with several basic factors. For example, we find that past asset values are significant in explaining future asset values. Importantly, this finding is not inconsistent with the idea that beliefs about future returns to land determine land prices. Rather, we might conclude that past asset values relate to current beliefs about future returns.

Prices in real estate markets swing dramatically. And often, after the market recoils from an extreme, we are hard pressed to justify its occurrence. However, even the most perceptive investors can only guess about the future, and it is likely that the process of formulating expectations has a great deal to do with swings in the prices of land.

Causality

The primary causes of asset values are past asset values and returns to assets. Real interest rates constitute only an indirect cause--to the extent that interest rates cause returns to assets. Actually, the relationship follows from the fact that the general economic conditions that cause real interest rates to be high cause returns to farm assets to be low. The converse is also true.

It is important that asset values cause asset values. In essence, this is to say that something happens in the market for assets that is not explained by past income and past real interest rates, but can be explained by asset values themselves or by market activity. This lends support to those who believe that, to some degree, people pay more for land on an upswing because land has recently gone up in value. Similarly, land tends to fall in value when it has been falling. Maybe this happens because people respond to other buyer activity rather than taking a careful look at the value of the land parcel in question. That is, knowing that other buyers have been paying more for land than had been the case earlier, the buyer does not hesitate to accept a higher price even though careful attention to returns to assets and so on might suggest a more cautious approach.

Impulse Response

Using the Featherstone and Baker model to study the dynamic response of asset values to external shocks, to interest rates, or to returns proves quite interesting. In essence, using the estimated set of parameters we can simulate the response of the system to a sudden shock to each of the variables in turn. In general, these results show little response of interest rates to shocks in other variables, much as one would expect. Real returns are not especially responsive except to real interest rates. The interesting results are the magnitude and length of asset value response to shocks in real interest rates and to real returns to assets.

In response to a one standard deviation increase in real returns to assets, asset values increase by roughly half a standard deviation in asset values for the first two years. However, asset values continue to increase for five more years, even though income falls back to the original level in three years. Asset values peak at one and a half standard deviations, and it takes about 13 years for the effects of a one year shock in income to completely diminish.

The response of asset values to a one standard deviation increase in real interest rates is nearly as great. After three years, asset values fall one standard deviation and recovery takes eleven years.

To get an idea of the frequency of shocks in returns or real interest rates in the past, one can examine the residuals of the estimated equations. Over the seventy years between 1915-1984 there were 21 years with residuals in the interest rate equation greater than the 2.62 percent standard deviation. There were 19 years with residuals in the income equation greater than the standard deviation of $2.766 billion. The greatest shock in real interest was 3.4 times the standard deviation on the positive side in 1920. The greatest real income shock was a positive deviation, in 1973, of 4.9 times the standard deviation.

The dynamic response of assets to returns and to interest rates follows, in large part, from the response of asset values to asset values. After an exogenous increase in asset values of one standard deviation, asset values take ten years to return to whatever starting point we have defined. Interestingly, asset values during years two through five exceed one standard deviation.

These results seem to support a hypothesis about farm real estate market dynamics: the basic farm problem of land boom-bust is in significant part supported by factors inherent in the market, such as the formation of beliefs about the future, and not by the past fundamental return and interest rate factors. This leads us to wonder about the degree to which the fundamental factors effect asset values.

Institutional Reorganization to Counter Concentrations of the Impact from Changing Land Values

Traditionally, when a market system performs in an unsatisfactory way, policy makers respond in one of two ways. Either they alter market conditions, or they treat the consequences. And these define the range of responses to the difficulties which land boom-busts visit on our farmers--now and in past cases. Yet a third alternative, which seems to us greatly preferable, might be to develop new institutions or organizational arrangements which allow existing market forces to perform more effectively in serving society's needs and desires. Price support and supply control programs are clearly attempts at altering market conditions. Income support payments are wealth transfers and represent efforts to treat the consequences of market outcomes. Unfortunately, each of these approaches can have unintended and undesired effects. Many farmers, and other observers of the agricultural scene, find serious fault with these attempted remedies because of those side effects. In fact, as we see how government programs and other policy manifestations con- tribute to the farmers' problems, we come to favor, more and more, the third possibility--institutional innovation.

Institutional Factors Behind the Problem

Farm Sector Financing

American agriculture has not developed, to any sig- nificant extent, institutions for providing outside equity financing. The vast majority of farmers operate as sole

117

proprietors, who depend largely on internally generated equity or debt financing. The typical pattern calls for a farmer to carry more debt during two periods of his life-- when he first starts out and then in middle life when he expands to capture size economies. Thus, debt has become a major factor in farm finance throughout the sector. But, this pattern also leads to a significant variation among individual farmers concerning how much debt they are using.

This patterning has led to a high saving rate among U.S. farm families and substantial accumulation of wealth in the sector as a whole. Sector wide that seems a healthy sign. But the uneven distribution of debt creates problems. And national policies have played an important role in this outcome.

Much of the new borrowing in the sector is by new or growing proprietors to purchase assets, mostly land, from people who are cutting back or discontinuing their operations. Sellers may use some of that money to retire their own debt, but most often they shift that equity into non-farm investments. Ultimately, a portion of what remains of a farmer's agricultural assets pass on to heirs. And those who no longer farm typically sell them to remaining farm sector proprietors. This process maintains the uneven distribution of debt among farmers.

It is important to note that the net outflow of equity capital from the sector is typically greater than the net inflow of loan funds. If this equity remained in the sector--as is the case in many European countries, because of laws, policies, and customs--there would be no great disparity of debt. But if we follow the European model, there would be a landed class with tenant operators or a class of owner-operators with hereditary wealth. Our society has chosen a different ownership system. We prefer our system; and, to ensure its survival, we have employed national policies of several kinds--not the least of which is the highly developed credit system for farmers and government programs to stabilize returns.

Government Programs

Government agencies considered many remedial programs during the 1920s and early 1930s. Pressure for farm relief continued until the government assumed responsibility for farm prices. Price and income support policies, as well as acreage adjustment programs, were initially

118

begun in order to raise cash inflows. Economists later argued that, among other benefits, the policies stabilized prices and reduced business risk. Agricultural credit programs were also initiated during the 1930s to address the capital needs of farmers. These policies continued because it was politically difficult to terminate them, but at the same time it was felt that these programs were consistent with the socially desired system of widespread ownership and control of farm assets. The farm sector has continued to develop under these programs and the programs have influenced that development.

Agricultural Credit: The United States has one of the world's most highly developed systems for injecting debt capital into agriculture. Several of the components have been overtly encouraged by the Federal government, among them the Cooperative Farm Credit System (FCS), the Farmers Home Administration (FmHA), and the Commodity Credit Corporation (CCC). Tax incentives for seller financing of farmland represent another method by which debt financing has been encouraged. Lest there be any doubt, a recent USDA bulletin (1984) amply documents the degree of Federal involvement in the creation of these institutions.

In addition to Federal involvement there are numerous cases in which states have assisted in boosting the amount of credit available to agriculture. Of primary note-worthiness are state banking regulations. Specifically, many states have, or have had, unit banking or limited branch banking to encourage local lending. In many rural communities this means more debt capital available to farmers.

Price and Income Support Programs: Returns were lower than expected during the Great Depression. There was considerable agreement that the cash flow problem of the time could be alleviated by raising returns and caus-ing them to be more stable. This approach seemed quite reasonable at the time, and it was consistent with soci-ety's overall objectives. However the logic is flawed.

Restating the approach demonstrates this. Price and income support policies work on the premise that address-ing problems brought about by errors in forecasting requires only that we change actual events in a way that validates the forecasts. A major difficulty, which is well known after fifty years of experience, is that this approach prevents prices and returns from guiding produc-tion and utilization as they do in a properly operating market. As a result, we must resort to other means of supply control and demand stimulation. That is bad enough, but the approach has other subtle, and perhaps even more important, effects.

Price and income support policies affect the structure of the farm sector in a manner which makes it even more subject to harm when asset values decline. When we contemplate the factors affecting a businessman's leverage decision, three factors dominate. These are the rate earned on investments, the variability of returns or risk, and the interest rate. One largely intended effect of public policy on these three factors has been to increase the amount of debt a rational farm businessman will choose.

In fact, a major objective of federal commodity programs covering such commodities as feed grains, wheat, cotton, tobacco, and dairy products, has been to stabilize returns, which allows farmers to rationally choose to have more debt.

Numerous government policies, ranging from FmHA programs to usury laws to agency status of the FCS have lowered the cost of borrowed money to farmers. When interest is lower, a farmer can afford more debt.

Public policy has not been totally idle with respect to rates of return either. While the intent of some commodity programs, such as dairy supports, has been to maintain profit margins, another effect has been to provide added incentive for farm expansion. A situation develops where the next or incremental investment continues to have a substantial expected rate of return. Furthermore, government sponsored research has assisted in pushing the economies of size further and further out along the size spectrum. Thus, many rational farmers in their initial and growth years will push leverage to the limit.

Financial Leverage Contributes to the Problem

Financial leverage increases income in good times, but leverage works in reverse in bad times. For the most part, use of leverage is restricted to levels at which the probability of survival is high for normal amounts of bad times. But, conditions are not always normal. As we have seen, the sector always has some participants with too much debt to survive a substantial downward revision in expected returns.

Downward revisions in beliefs about expected returns cause declines in the value of land and in the wealth of

120

farmers. At the same time, downward revisions in returns for farmers may mean that food prices are expected to be lowered and society as a whole is not worse off. Hence, the wealth declines represent transfers of wealth. Consumers gain, at the expense of the owners of land and farm assets. While gains to society are spread over a large population, the losses are more concentrated. Since much havoc, pain, and suffering is associated with land price busts, we need to understand the alternatives for alleviating the problem.

Institutional Innovations to Address the Problem

The idea that government policy need not be limited to changing market conditions or treating symptoms of problems is an important one. Of course, these approaches will often be appropriate or even necessary. But, they do not include the first alternative to be considered. The preferred alternative is development of institutions which use market forces to achieve desired ends. Markets themselves are institutional innovations which give us much individual choice and satisfaction. More specifically, money and banking, credit, futures exchanges, corporations, public utilities, and the judicial system, represent institutions which developed to solve problems which vexed the economy at one time. We have designed these institutions to make the market outcomes more efficient and more desirable. They seem to have accomplished that. While numerous farm policy makers have actually urged institutional innovation in the past their efforts have borne little fruit. Yet for some reason, the farm sector and those responsible for formulating farm policy have never emphasized this approach. We think the reason for that is that these people have gotten mired down in the intervention policies that dominate the scene because they have not understood that the issue was a structural one concerning the patterns of ownership and financing of farm assets. We think that this is an area which has been too little explored.

Summary

The structural organization of the farm sector results in chaotic and disruptive wealth transfers with much personal pain, whenever beliefs about future returns to farm assets are lowered. This problem points up a deficiency in our current system of ownership and financing of farm land. Substantial changes in beliefs about future returns have occurred a number of times in the past. Each have had similar disruptive effects.

In this chapter we have said that the changes in land value are transfers from land owners to consumers of agricultural products, and argued that these transfers are unavoidable long as the future cannot be forecast accurately. But, we believe that the difficulty stems from the concentration of losses among landowners and especially among debt holding landowners.

Considerable attention was given to the role which farm real estate plays as an adjustment mechanism. The adjustment process is hampered by the illiquidity of farm land. Evidence is cited that suggests shocks in returns to assets can set off boom-bust phenomena whose effects are felt for many years--i.e., that both the boom and the bust may feed upon themselves and overshoot eventual equilibrium asset values.

Farm sector financing, financial leverage and public policy were noted as factors behind the problem. And the idea was stressed that public policy need not be limited to changing market outcomes or treating market consequences in order to address the problem.

NOTES

1. We follow the convention that an asset earns economic rent whenever it earns returns greater than the minimum required to continue its employment in its particular function. That a large part of the return to land is of this sort was originally pointed out by Henry George in the nineteenth century.

2. Melichar's data from 1910-1984 provide the basis for our summary.

122

6

Some Policy Options

As we have developed our analyses, several truths
have emerged which should be obvious. Numerous American
farmers of all kinds and in all areas face severe financial stress. Their problem is real. We cannot argue it
away or dismiss it. The nature of this financial problem
is not what many analysts, observers, and policy makers
have thought it to be. It has little to do with commodity
prices or income distribution, which are only symptoms.
Rather, it stems from the basic structure of the farm
economy--more specifically, from our approach to land ownership and to our related ability to adjust to negative
changes in asset values. Crucial in all of this--the
value of a farmer's assets and the way he finances them
are at the heart of the problem. When a situation is this
serious and this widespread, it necessarily becomes a concern of policy makers at all levels--public and private.
Yet, policy approaches, concentrating as they do on prices
and income guarantees and ignoring modes of increasing
equity financing, seem to have exacerbated the problem
rather than to have moved our economic institutions toward
any kind of solution.

In some sense, what we say is not entirely novel. In
1948, John Timmons stressed the importance of the issues
we are addressing. He urged action. He said there was
widespread awareness of the risk involved in farm asset
ownership:

> Much has already been done by the colleges, Extension
> Services, Department of Agriculture, banker's organ-
> izations, and other groups to educate farm people and
> all prospective owners on the dangers of land booms

123

and heavy mortgage debt incurred in the purchasing of inflated lands (Timmons, _JFE_, 1948, p. 99).

But he feared that the lesson might be forgotten, that the problem would again need to be addressed.

Another important area of inquiry is concerned with getting ownership conditions ready for possible declines and wide fluctuation in farm income. Some of this research should be concerned with preventing excessive mortgage debts and unsatisfactory purchase arrangements. But an increasing amount of attention should be devoted to remedial measures to be undertaken if and when serious income fluctuations threaten the stability of farm ownership (p. 99).

And he called for policy research in anticipation of the problem which faces us today:

Now is the time to study and effect legislation to stabilize ownership conditions. Let's not wait until serious difficulties develop and the pressure of an emergency situation demand that speed characterize remedial action. Attention should be given to the problem of debt moratoria, debt adjustments, corporate holdings, and large holdings (p. 99).

We could use any number of policy approaches in this problem. Agricultural economist J. Carroll Bottum has observed the policy process since the last land boom-bust of the 1930s and he insists that any proposed alternatives can only be variants of others previously tried. Perhaps so. But we think a glum appraisal of past policy success may have dulled the profession's aggregate imagination. We think there are possibilities which extend beyond previous attempts. Indeed, the way we conceive the problem differs from that which guided past policy prescriptions. Policy makers of virtually every affiliation and persuasion have clung so steadfastly to these traditional policy approaches that, before we offer our own sense of the policy alternatives which might actually contribute to a solution of the problem, we need to consider why traditional approaches have done no more for farmers than they have. Only then do other ideas make sense, for only in that context can we motivate our search for departures from what is demonstrably unfulfilling.

Recasting the Problem--A Framework
for Problem Resolution

Throughout our discussion, we have presented a view of the farm economy and an analysis of the predicament which currently afflicts it. Certain details of our presentation depart from traditional policy views. For example, our claim that asset values, and not commodity prices nor income distribution, lie at the heart of the problem counters the statements of at least some public figures. So does our observation that many of those in the direst straights are among the best managers and the most to be treasured among our farmers. And while certain of these claims certainly depart from the testimony one most often hears in policy deliberations, we find ample evidence to support them--evidence which is not esoteric and which requires no special handling. In fact, we take it that the discussion so far has rendered many of these points intuitively obvious.

Given that, we assume concurrence on eleven central points. As a review, we restate these eleven points and briefly outline our thought concerning them. This provides a rational foundation for our discussion of traditional policies and for our suggestion about how we might best enhance the structure of our agricultural economy.

Some Fundamental Points

1. **Huge differences among individual farmers based on such factors as management skill, farm size, equity capital, income opportunities and sources, crop and enterprise mix, and land quality narrowly constrain any generalizations about the sector and carry important implications for policy formation.**

An important observation from study of various kinds and sizes of Midwestern farms (see Chapter IV) concerns the extreme variability among farms. Though important, measures of farm size, financial leverage, and kind of farm leave unexplained a great deal of variation in profitability, financial stress, and general well-being. The fact is that some individuals in any class do well, while others who appear to have good situations do poorly. Accordingly, we cannot expect programs which attempt to address sector wide areas of difficulty to solve all the problems. We must realize that whatever the policy, some farmers will continue to do poorly.

The observation also has implications for the kinds of policies which are most likely to be effective. Policies which present new, non-mandatory opportunities with minimal market interference could allow some individuals to do better without harming anyone else. So we prefer institutional developments which improve market operations and lessen transaction costs to policies which attempt to control market outcomes.

2. <u>Farm income is not the farm problem</u>.

Livestock production can be profitable for output prices far below current ones if the price of corn and other feedstuffs is low enough. Corn production can be profitable at price levels far below those currently observed if the price of land is sufficiently low. Melichar (1984) shows that real dollar returns to farm assets were higher than prior to the boom period. Yet the problems were severe.

Expectations for future returns increased during the seventies and have since declined. Various analysts have demonstrated that land price adjusts, rather quickly in fact, to reflect current and expected future returns from crop production (for example see Dobbins, <u>et al</u>.). Land price is a residual element, determined by market conditions and serving as the final adjustment factor which allows overall equilibrium. The farm problem arises when downward land price adjustments are necessary and the existing ownership system does not facilitate that adjustment.

3. <u>Our system for financing agriculture makes success in large degree dependant upon the accident of when one is born</u>.

Substantial debt financing is normal in our system of farm ownership. Typically, the cost of that debt derives from the amortization of the land value at a single time. As a result, the accuracy of the income forecast for the land at that point crucially determines the lifetime success of many farmers.

A consideration of the predicament of farmers in the 1980s suggests that all owners of farm assets have sustained losses since the peak of the land boom, but the degree of stress for an individual depends upon when he made the initial purchase. Consider two land owning farmers, one who bought before the recent land boom and one at the peak of the boom period. If real returns hold at recent levels, asset values will stabilize above their real values for the 1950s and 60s. In that case, the first farmer will have made overall gains. This

126

individual made large "paper" gains during the seventies and then somewhat smaller, although sizeable, "paper" losses during the early eighties. The "paper" gains represented opportunities for cash gain during the land boom period, only some of which remain available today. In a sense, this first farmer suffered only a lost opportunity as a result of the recent decline. The remaining "paper" profits provide some consolation. The second farmer, who bought assets at their peak values, has sustained substantial "paper" losses. Currently anticipated income streams are substantially less than those expected when he made the purchase. This farmer's net worth has declined, and he has no previous gain in net worth to console him. These cases illustrate the speculative element in asset ownership which results from the uncertainty about future returns. Clearly prospective owners must be capable of handling the possible losses or of facing the possibility of failure, however remote.

4. <u>At any moment in history, a substantial number of farmers have monetary incentives to expand</u>.

There is often substantial incentive to expand in order to use management ability fully and to increase the total returns to management and operator labor. The course of technological progress, which changes the economies of size, combines with growing capital requirements, income variability and the struggle to retain earnings in the form of equity in the sector to perpetuate this situation.

5. <u>Farmers take a rational approach to growth strategies, weighing potential gains against associated risks</u>.

When farmers finance land purchases with debt, they consider whether they can handle payments out of cash flow during periods of low income. They take into account their current financial position when they contemplate buying land at high prices. And they weigh these factors against potential gains from expansion and land ownership. Lower real interest rates, increased returns, and a long period of relatively stable growth tilt the strategy toward expansion.

Schultz considers ways in which the private market carries business risk in agriculture, and discusses an important adaptation on the part of the farm firms together with a major drawback:

In principle, this solution is simple and straightforward. What is needed is a family farm with enough assets so distributed that it can cope satisfactorily

127

with the fluctuations in income caused by variations in yields. Both the interest of the farm business and of the farm household are involved. Two major difficulties arise in practice: The amount of assets required is often too large for a family farm to finance under existing conditions; and most farm families, even if they had sufficient assets, have relatively little experience and information on how to hold such assets in order to manage their availability when confronted by yield instability (Schultz, 1953, p. 333).

Schultz presented a specific example, an 800 acre wheat and livestock farm on the northern plains which required family equity of $100,000 in 1953 dollars (3/4 in real and 1/4 in cash assets) in order to maintain a high probability of survival.

Gabriel and Baker (1980) have developed a theoretical framework to explain how farmers can adjust total income risk by varying the level of financial leverage. Their framework provides a rigorous description of the principle to which Schultz referred.

6. <u>Because returns to land ownership are uncertain, the investment has a substantial speculative element, and financial leverage magnifies the uncertainty</u>.

A substantial number of farmers want to expand when there is an abundance of retained earnings in the sector and a general belief that returns will continue to grow. Then land booms occur. When people generally agree that that forecast was in error, then the bust comes.

Financial leverage magnifies the speculative nature of asset ownership. By increasing the proportion of assets financed with fixed rate debt obligations, a farmer can increase his opportunity for profit or loss. If the rate of return on assets is greater than the rate of interest on debt, then the rate of return on equity is increased. On the other hand, if the rate of return on assets is low enough relative to the rate of interest and the degree of leverage, then the rate of return on equity can be negative. So, a farmer holding land through the boom period to the present may have multiplied his initial net worth many times using debt. Another farmer may have multiplied his loss many times by using debt to finance purchases at peak asset prices.

However, a farmer who uses financial investments with fixed rates of return can reduce the speculative nature of asset ownership. In fact, he can eliminate risk. He simply reduces his debt and invests a portion of his equity in interest-bearing instruments--negative debt. Holding all of the equity in risk-free investments eliminates risk. Most people prefer a mix of debt and equity financing. The effect is similar to diversifying an investment portfolio.

But, while diversification can reduce an individual's risk nothing can reduce the uncertainty of returns to land ownership. Future returns to farm assets are unknowable ahead of time. Given that, there is no way around the fact that asset values are subject to change as beliefs about future returns change.

7. For society in general, changes in the value of output from farm assets, such as food, offset changes in the value of future returns to assets.

When the expected future returns to farm assets rise, farmland owners gain but the expected future value of consumer food expenditures rises by a like amount--so consumers lose--the converse is also true. The value changes are transfers and neither net welfare gains nor net welfare losses.

However, there is great disparity in the magnitude of the effects on individuals. Lower food costs for the entire society effect the many in a scarcely noticeable way. But losses in asset values are concentrated in the form of lower asset values among the small percentage of people who own land. Furthermore, the perceived pain is greater for those owners using debt financing than it is for those using equity financing.

8. The argument that education is an eminently reasonable preventative which seems to wear off prior to each event is falacious.

> Education, it is claimed by some authorities, is sufficient protection from land inflation and other dangers. According to the argument advanced, the important task is to point out the dangers, make all potential buyers conscious of the hazards and urge them to proceed cautiously in periods like the present when income is unusually high. As part of the education program, farm operators are urged to get out of debt and stay out of debt. Another phase of the campaign is a solemn warning of what may happen if land is bought on credit during periods of high income. The final lesson to strengthen the willpower

129

of the farmer is a grim recital of the tragic events which followed in the wake of some particular land purchase during the last boom (Murray, 1943, p. 207).

To suppose that education "wears off" is inconsistent with an assumption of rational behavior. The idea that people ignore certain available information is inconsistent with our most fundamental understanding of behavior.

Another view is that farmers are reasonably aware of the problems involved with debt financing and act with due caution. The problem arises because farmers have perceived the value of potential gains as exceeding the value of possible losses.

9. <u>Farm problems are a direct result of the existing structure of agriculture</u>.

We have based the current economic structure of agriculture upon private property rights and use uncertain profits and losses to guide entrepreneurial endeavor. We traditionally resist the accumulation of family wealth across generations and, furthermore, feel that family wealth should be distributed equally among the children rather than concentrated with one member who remains on the farm. We have, in effect, resisted the means by which Europeans maintained equity financing for agriculture before colonization. Americans chose a different approach. Whether by design or accident, an extensive agricultural credit system has developed. Debt financing plays an important role in United States agriculture.

The problem is that debt financing is not equivalent to equity financing. Adjustments in the face of downward revisions in beliefs about future income returns differ with the source of financing. Where debt is the primary source of financing, assets must be liquidated when income is expected to be insufficient to meet obligations on the debt. There are other options available when equity represents the primary source of financing. Acknowledgement of lower expected income does not necessarily entail the sale of assets financed with equity. Thus, liquidity risk is greater as debt is greater. A system based upon debt financing is simply less resilient than one based upon equity financing.

10. <u>Given the current economic organization of the farm sector, land boom-busts are disruptive events which cause chaotic wealth transfers which involve allocative inefficiencies and much personal pain</u>.

The aftermath of the current land boom is overinvestment in land and depreciable assets. Farmers paid excessive prices and used too much debt financing because of their expectations of steadily growing farm income. Because those expectations have proved false, farmers throughout the sector have suffered huge capital losses. Many were not prepared to deal with losses of the magnitudes of those which have come to pass. Many had gambled on magnifying potential profits through aggressive use of financial leverage. For these individuals the outcome has been magnified losses.   -

In cases where the assets were primarily financed with equity, the loss is relatively easily born. But for many, the losses will cause severe suffering. Where substantial debt is involved, the losses will in all likelihood dictate reorganizing the capital structure of the farm. Many farmers will need to decrease debt relative to income returns and asset values, most commonly through downsizing the operation or selling assets and leasing them back. In many cases debt was high enough that the decline in asset values together with recent cash short-falls has completely eroded the equity capital. A steady reduction of the recoverable portion of debt has begun. The creditor is sharing in the loss in those cases where equity has been completely eroded and, depending upon contract terms, may be called upon to share in losses in other cases as well.

11. **The development of more equity financing in American agriculture would do much to ameliorate the financial problems of farmers.**

The farm sector might increase equity financing to reduce liquidity risk in either of two ways--by reducing the drain of equity from the sector, or by increasing the flow of equity into the sector from outside sources. Establishing more family corporations could reduce the drain on farm sector equity attributable to discontinuing proprietors and offspring who do not choose to remain in farming. Public corporations could provide the means for attracting outside equity financing. Each approach would result in greater resiliency in the farm sector.

Policy Options

The goal of any policy we adopt should be to make society better off in some sense. We all like to see improvements in the general welfare through policies which improve the lot of some people without leaving anyone worse off. Of course, this is an idealization.

Legitimate as those broad goals are, the imaginable policy options are so numerous, and their effects so diverse, that we cannot hope to compare them usefully unless we establish a coherent framework which will allow study of their relative merit. At the very least, such a framework should help us determine the source of the policy benefit --from what or whom the gains derive, whether the policy intends to produce short term or long term gains, and whether the policy has been tried before. A policy which makes the economic system more efficient adds to the productivity of society and is indeed beneficial. Other policies attempt to transfer existing wealth according to the society's notions about how to increase overall welfare. Many traditional farm income support programs involve just such transfers. The deficiency payment is conceptually a program to accomplish transfers from taxpayers to farmers and owners of agricultural assets. Price support programs, as implemented in the past, have had both kinds of characteristics. They removed some of the downside risk from price expectations. And they made the system more efficient as long as stock accumulation was not costly. But, if grain purchased by the Commodity Credit Corporation under the auspices of price support must be sold at a loss to taxpayers or given back to farmers as an incentive for supply control, the effect is much the same as a deficiency payment.

Another useful dimension for studying the comparative merits of policy options arrays them according to whether they produce immediate or long term gains. Some policies are of a temporary nature and some are expected to provide permanent service. The recent Payment-In-Kind program was intended as a temporary, quick fix solution to both the financial problems of farmers and stock accumulation problems associated with other policies. On the other hand, bankruptcy law and the commodity futures exchange system are results of policies which have provided permanent benefits.

A further dimension for classifying policy options involves whether the policy has been used in the past in this or other countries, whether it is currently being considered, or whether it represents a new policy approach. For example, price and income stabilization schemes have a long history in the United States. Initially instituted in response to the financial crisis of the Great Depression, our government has maintained them ever since. Strict restrictions on farm debt obligations were used in Switzerland as a long term solution to problems of too much debt. We propose a new policy to facilitate bringing equity into the farm sector from outside sources and to provide additional liquidity in the market for farm real estate.

A Framework

At this point, we are not focusing so much on particular policies as on classes of policy alternatives. Within each class, several variations are imaginable. Yet all the variants will, in the end, derive from similar sources and produce similar benefits. Table 6.1 outlines these classes of alternatives. It allows us, further, to distinguish between classes of policies which provide short-term aid during bad times and those which produce long-term benefits through more efficient provision of services.

The first alternatives the table lists involve commodity policies and credit programs and institutions--all aspects of the policy repertoire which have long histories in U.S. agriculture. However, because these alternatives use aspects of the agricultural credit system which increase the economic incentives to use more debt financing, none has even the potential to alleviate the problem on which we focus. Indeed, measures like subsidized interest rates only worsen farmers' problems.

Bankruptcy law, in contrast, provides genuine short-term aid on a case by case basis. This law consists of a set of rules through which a farmer can work out problems of financial maladjustment of the kind so likely to occur in the aftermath of a land boom. Basically, the court mediates between debtor and creditor, as well as among creditors. In providing short-term aid during a financial crisis, the law allows the market to work more smoothly than it otherwise might. In sum, the bankruptcy law attempts to alleviate the farm problem by providing a structure through which individual farmers can resolve their debt problems.

European countries have devised a variety of land ownership patterns and legal devices which, in effect, discourage excessive debt financing.

Shortly after the Great Depression, Switzerland enacted legislation which limits people's rights to assume too much debt. Besides severely limiting the farmer's right to contract for debt, this law provides for reductions in his obligations on debt owed to family members

133

Table 6.1. Framework for Comparison of Policy Options.

	Short-Term Aid in Financial Crisis	Long-Term Benefit to Farm Sector	Historical Experience to Farm Sector	Potential to Aid Farm Problems	Source of Improvement	Federal Government Adminis-trative Burden	Budget Outlay	Welfare Effects Rest of World	United States Consumer	United States Taxpayer	Farm Sector
No Policy Action	None	?	?	Little	-	High	High	-	-	-	-
Commodity Policies:											
Non-Recourse Loans	Some	?	Much	Bad	Transfer	-	Moderate	Good	-	Bad	Good
Deficiency Payments	Some	?	Much	Bad	Transfer	-	High	Good	-	Bad	Good
Supply Control	Some	?	Much	?	Transfer	Medium	High	Good	Bad	Bad	?
Farm Credit Institutions:											
Federal Land Bank	?	Yes	Much	None	Efficiency	-	-				Good
Production Credit	?	Yes	Much	None	Efficiency	-	-				
Farm Home Administration	?	?	Much	Bad	Transfer	Medium	-			Bad	Good
Farm Credit Programs:											
Debt Moratoria	?	No	Slight	Bad ?	-	-	-			Bad	Bad
Subsidized Credit	Some	?	Some	Bad	Transfer	Moderate	Moderate			Bad	?
Loan Guarantees	None	No	Some	Good	Trasfer	Moderate	Moderate			Bad	Bad
Law and Structure:											
Bankruptcy Law	Much	Yes	Much	Some	Efficiency	-	-				Good
Private Landowning Corps	Slight	Yes	Slight	Good	Efficiency	-	-				Good
Family Landowning Corps	?	Yes	Slight	Good	Efficiency	-	-				Good

134

Table 6.1 (continued).

	Short-Term Financial Aid in Crisis	Long-Term Benefit to Farm Sector	Potential to Aid Farm Problems	Historical Experience	Source of Improvement	Federal Government Adminis-trative Burden	Federal Government Budget Outlay	Welfare Effects Rest of World	Welfare Effects United States Consumer	Taxpayer	Farm Sector
Other Countries:											
Few Hereditary, Wealthy Landowners	Much	?	Good	Eng.	?	-	-				?
Restrictions on Debt	Some	?	Good	Switz.	?	Low	-				?
Bail Outs:											
Restore "Boom" Condition	Much	?	Bad	None	Transfer	?	?		Bad		Bad
Extend Additional Credit	?	No	?	Some	?	High	High		Bad		Good
Forgive the Debt	Much	?	Bad	None	Transfer	Low	High		Bad		Good
New Proposals:											
Interim Land Ownership & Financing (Harl)	Much	No	No	Similar to 1930s programs	Potential Efficiency	High	Medium		Good		Good
Publically Administrated Private Landowning Corp. w/Futures Trading on Stock	During startup Yes	Yes	Good	Yes	Efficiency	Low	Low		Good		Good

during periods of low farm income. While that provides useful short-term aid, the long-term benefits are doubtful because, though such measures prevent one kind of problem, they also appear to limit the ability of the farm sector to expand quickly in good times.

Other European countries have taken another approach to avoid overuse of debt financing. They promote land ownership stability through their systems of hereditary land ownership. This system certainly eliminates the drain on farm sector equity that results from individual farmers leaving the sector. Debt financing plays much less a role than in the U.S., so that problem is greatly reduced. However, this ownership structure clashes with the strong American beliefs in widespread ownership of farm assets and open opportunities for new people to enter the sector.

Two sub-categories of the "bailout" option could solve the current problem of excessive debt but are otherwise unsatisfactory. A third bailout ploy is illogical on its face. It seems absurd to think that government guarantees and extension of more debt can even hope to provide a long-run solution to too much debt. The other bailout solutions are at least logical, though they leave much to be desired in other ways.

Some logically possible approaches involve restoring the conditions which farmers, and others, expected when they contracted their high levels of debt. There are at least three ways of doing that.

One method would be to increase the capitalized value of the future stream of returns to assets through a new round of unexpected inflation--an approach Warren and Pearson, among others, strongly supported during the 1920s and '30s. Doing this might relieve the current financial crisis in the farm sector but it would surely disrupt other sectors of the economy. The current social mood rejects further inflation.

A second alternative would be to increase the value of farm assets by issuing grants to farmers in the amount of the excess debt. A one-time infusion of $200 billion, say, could restore the balance between debt and equity.

A third option would approach liabilities instead of assets and simply forgive the excess debt. That would solve the current crisis by converting debt into equity, in effect, and restoring the balance.

All of these options are fraught with difficulty. Their feasibility suffers from the vast burdens they place on taxpayers or lenders. Taxpayer financed programs would have high administrative costs, would inevitably involve payments to some for whom the program is not intended, and may well introduce permanently high levels of government involvement in farming. In essence, it is clear that restoring conditions expected during the 1970s and forgiving substantial amounts of debt are not feasible options on their own merit. The further fact that the federal budget is in a state of enormous imbalance makes the bail-out options seem pragmatically without feasibility as well.

Along with these familiar options, Table 6.1 includes two new proposals. Neil Harl, an Iowa State University economist, has suggested a system of temporary government land ownership. That has merit since it provides aid during the period of financial adjustment at only moderate cost, given Harl's estimates. However, since Harl's proposal does not address the basic problem--the condition which is responsible for the financial crisis, it offers little chance of long-term correction.

The final category in Table 6.1 outlines our proposal which involves developing a new institutional structure to draw outside equity financing into the farm sector. To the extent that this device would increase liquidity, and so the price for farm land, it would provide short-term aid to those suffering during the present crisis. Because this approach would allow farmers to trade in land futures in an organized way, analogous to commodity futures trading, it would help farmers minimize the results of broad declines in returns to farm assets. Also, this device would replace volatile debt financing with more stable equity financing to further the long-term stability of the farm economy.

This framework, then, allows us to focus attention on relevant details of the various kinds of policy options so we can compare them in principled ways and evaluate what kind of results they are likely to produce. We can identify traditional commodity and credit programs as simple transfers, from either consumers or taxpayers. In fact, programs like these produce, at most, an illusion of benefit. Often they exacerbate the problem. Bankruptcy law and the corporate structure of business provide actual benefits by lowering transaction costs involved in market operations, and making the market more efficient.

We can see, through the use of this framework, whether a policy approach will produce long-term or short-term benefits to the farm sector. This distinction is especially important because the two kinds of benefit may conflict. The best short-term policy may be the worst in terms of long range results. We can also see that gain from improved market operation is preferable to the subjective gain that comes from a transfer. Finally, we can see that some kinds of policy cannot, even in principle, address the central economic problem which has to do with the nature of farm financing and changes in the value of farm assets--especially land.

Alternatives and Consequences

The classification set forth in Table 6.1 delineates imaginable policy approaches. And it provides a framework for considering possible outcomes. A logical next step is to consider some actual kinds of policy proposals which have received public attention in recent times. Along with identifying those, we offer qualitative evaluations of their possible consequences. Actual policy proposals include:

1) a market solution to the farm financial crisis with no substantive change from current policies,
2) enhancing benefits of our traditional price and income support policies during "bad" times,
3) debt moratoria and loan guarantees,
4) legislation to limit the rights of farmers to contract "excess" debt in the future,
5) maintaining a supply of government owned land as a buffer stock,
6) using local committees as counsels to ease financial stress and to work out individual problem solutions.
7) Neil Harl's proposal for interim land ownership and financing by a federally chartered entity,
8) a futures market in land to aid in transferring the risk of changing levels of farm income,

This list does not achieve a one-to-one correspondence with the possibilities of Table 6.1. Yet, these policy suggestions capture the range of public discussion and show how our framework can serve to advance our thinking about this vexing public issue.

Alternative 1: No major changes in agricultural policy.

The market solution to the current financial crisis will be a broad mix of forebearance and foreclosures.

138

Forebearance will take the form of extensions and compositions which debtors and creditors work out individually. Left alone, the market will adjust. Given this solution many farmers will experience severe pain, and the solution will not be perfect. But neither will any known alternative be perfect.

We can think of several typical situations which will be likely to result from an absence of major policy changes. If we suppose that returns will stabilize at their recent levels and that asset values have fallen to the point that they are now near their new long-run level of equilibrium. Many farmers are in an extremely weak financial position--many are technically insolvent. But that, by itself, does not imply that they face foreclosure or other final solutions.

For example, if a farmer is technically insolvent because of declining asset values, financial advisors would not necessarily push him towards disinvestment. For one thing, the recoverable values of these assets may have declined to the point that a sufficient rate of return is possible with the asset in the hands of its current owner. Lenders may anticipate that the current owner can earn returns on the asset equal to or higher than any other potential operator. For what is important in this consideration is the expected rate of return to the current market value of the asset and not to the value at the time of purchase. This is true in spite of the fact it is the later value against which the debt was contracted.

Further, if we assume that the intent of the creditor is to recoup as much of the loan as possible, it follows that he wants both asset and borrower to earn the highest possible returns. And the present owner may offer the best chance of achieving that. At the same time, it seems reasonable to assume that the borrower also wants to make the best of the situation and so wants to earn the highest possible returns. In most cases, unless he could earn more in another line of business, that means continuing to farm. Given that, the best alternative for both lender and borrower may be to continue the relationship, though it is quite likely that they will need to renegotiate the loan to reflect the new, and lower, rate of returns.

In some of those cases, the debtor may find it simply impossible to service the debt. Others, perhaps those less highly leveraged, may find debt service possible but markedly unpleasant. Perhaps both sides will have to acknowledge the fact of loss and agree to share the consequences. In a few more extreme cases, the remaining balance due may be so much more than current market values that the debtor may see defaulting on the mortgage as a viable option. Clearly, if the creditor's best option is for that farmer to maintain operations, there is room for negotiation in this kind of situation.

Many cases will be less clear cut. Suppose that a farmer is solvent but has lost a substantial portion of his equity because of an ill-timed purchase of land at, or near, the peak of the land boom. His lender may consider him a poor risk given the variability of his income. He may have too small a probability of survival given the current size of his operation in relation to his remaining equity. Both he and the lender may decide to downsize his operation in light of the new equity situation. It is more likely that the lender will make this determination independently. Too often contractual arrangements leave the borrower little room for negotiation. The lender may force the farmer to downsize in order to lessen the loan risk in spite of economies of size and other efficiency considerations. Even though those are matters which call for thorough discussion, communication between the parties is difficult in these cases. Too often one or both parties are poorly prepared to deal with the realities of the loss and its implications. Communication is more difficult in the real world because seldom is either side certain about the actual magnitude of the losses.

In fact, our general uncertainty and imperfect knowledge about these matters often prevents our recognition of declines in net worth. While that seems like it ought to be a relatively straightforward matter, it seldom is. We do not immediately notice reductions in asset values. At first, we may well recognize that returns have not met expectations, yet we may attribute that to temporary conditions--weather or the failure of a treaty with a grain buying nation. And even if our analysis of what we can anticipate is on the sober side, the positive transaction costs which accompany such remedies as disinvesting or reorganizing may make continuing as is seem the best option. That could be true even in a period when income does not cover interest on debt.

If, in fact, the lower returns result from temporary causes, using additional debt to cover income shortfalls constitutes a reasonable strategy. However, if the causes of the lower returns are not actually due to temporary conditions, but the result of something more fundamental, the additional debt strategy can exact a tragic cost.

Each year a farmer makes up cash shortfalls through more borrowing, he increases the likelihood of shortfalls in ensuing years. Also, if the causes were not really temporary, then asset values have continued to fall. Clearly, our market assigns a premium value to accurate perceptions about matters like these.

It also seems obvious that farmers whose level of debt is high enough to cause serious problems in the event of cash shortfalls should probably disinvest before the value of their assets erodes completely. Those who take the longest to recognize the new conditions will face a double-edged problem. Besides suffering greater losses from the decline in the value of their assets, they are likely to have acquired too much of debt in their efforts to cover cash shortfalls.

The necessary strategic decision-making process is complex. Errors are expensive. And every aspect of the situation is unpleasant. Communication between lender and borrower is extremely important though likely to be painful as thousands of farmers work out these problems across the country.

In the long run, as proponents of doing nothing about farm policy claim, a market oriented policy which significantly reduced the level of price supports would increase farmers' sense of farm business risk. That, in turn, would discourage the use of debt financing. Taken on its face, that seems a desirable result, for debt is certainly a major component of the problem. However, this policy option is silent about a source of financing to replace the sector's reliance on debt. Farmers would still need to finance their operations, so doing nothing will not be the answer.

As consequences of making no major changes in farm policy, we might expect

- the market system to bring about needed adjustments in capital structures on an individual basis
- this solution to be painful for many as both debtors and creditors struggle to recognize and cut losses because, in part, of the difficulty of communication and negotiation and also because either party may well be ill-prepared to deal with the realities of the loss
- a tendency for lenders to underestimate the future potential of some farmers because of the bleakness of their present financial situation

- this adjustment to proceed slowly and, very possibly, to overshoot the needed correction as a pessimistic outlook replaces the earlier optimistic one
- education concerning the costs of over-leveraging and over-investing to intensify and to extend the time between this crisis and the next
- the rural sector to continue--in spite of the fact that many farmers and lenders have been virtually wiped out--though with a conservative shift in attitudes about consumption, saving, financial leverage, and investment.

Alternative 2: Enhancing a major policy approach utilized in the past--income and price supports.

A typical approach to solving the farm financial problem involves income and price supports. Although these policies take many forms, they basically raise expected future returns from farming and from the ownership of farm assets. Since asset values ultimately derive from returns, this kind of policy serves to sustain those values. Or, as one USDA publication reports, "since the benefits are proportional to the amount of production, they tend to be capitalized into the value of the more limiting resource, land" (USDA, p. 104). According to an estimate by Reinsel and Krenz, the total capitalized value of farm program benefits amounted to 8 percent of the value of farm real estate in 1970.

This kind of policy definitely produces benefits. The land owners of record when the policy is announced certainly gain. And except to the degree that a larger initial investment in land is required, it would not be detrimental to new entrants into the sector.

The problem is that such a policy, once begun, is rather like having a tiger by the tail. As long as it continues, farmers, and lenders, receive aid from it during a bust period. But income and asset levels will drop from their artificial levels the moment the policy maker announces its termination. And landowners who hold assets throughout the term of the policy would then bear a cost which would roughly offset the benefit they derived from the introduction of the policy. People who entered farming during the policy's active period, purchasing land at supported levels, can really suffer severe damage upon termination of such a policy. People who hold land at the outset of the policy and then sell out during the term of the policy can gain significantly.

142

Furthermore, this kind of policy does nothing to relieve the actual farm finance problem. Forecasts of future returns continue subject to the usual error. Land booms continue to lurk as a potential problem. Worse, in treating symptoms of the fundamental problem, rather than the problem itself, these policies tend to impede the market's effort to solve its problems by sustaining asset values just when market forces are trying to push them downward. That kind of counterforce in the market may also obscure the communication process, problematic at best, on which farmers and bankers base their decisions.

Ultimately, this kind of policy which reduces business risk has the effect of encouraging the use of debt financing. In the long run, that renders the farm economy significantly less resilient, less well-equipped to deal with the vicissitudes of the marketplace. Accordingly, while such a policy might create some short-term benefits, its long-term effect is to worsen farmers' economic problems in the future.

So, as consequences of enhancing traditional income and price support policies, we might expect

- farmers of record at the outset of the program to gain by the capitalized value of the expected stream benefits
- the benefit stream to be capitalized into the price of land
- new entrants to suffer no loss from this policy--except to the extent that it requires larger initial investment in land or rent
- ultimate policy termination to impose a cost on landowners roughly equivalent to the initial benefits
- this policy to impede downward adjustments in land values which income forecasts would otherwise motivate
- a high expenditure level to maintain prices and incomes at the levels which the peak boom years established
- even very large expenditures to leave gaps in coverage, especially among those who purchased all of their holdings at peak prices
- this policy to increase the use of debt financing and to reduce the resiliency of the sector in the face of future downturns.

Alternative 3: Debt moratoria and loan guarantees.

Public outcries for moratoriums on farm foreclosures, widespread during the Great Depression, are popular reactions to the current farm financial crisis. Unfortunately, neither debt moratoria nor loan guarantees represent positive approaches to the problem. Where the problem is too much debt in relation to income generating power, we must adjust the capital structure so it corresponds to the reality of the marketplace. Waiting to adjust simply makes the adjustment more difficult.

The Chapter 11 section of United States Bankruptcy Code provides legal devices which guarantee debtors time to structure a plan for reorganization. Such a measure-- which enforces communication between debtor and creditor, careful thought about the overall financial situation, and hard choices about courses of action--is more likely to produce genuine benefits in that it necessarily deals with only one case at a time. An across-the-board moratorium actually could worsen the situation for those who really need to reduce debt by selling assets, or otherwise restructuring their business, and who would do nothing because of a moratorium. To the extent a moratorium encourages those people to be uncommunicative about their financial affairs, it performs a major disservice to the farm community. For the interest liabilities of those in trouble will continue to compound, and their remaining equity will erode further. Similarly, lenders losses would continue to mount.

The debt restructuring program announced on September 18, 1984 and implemented in early 1985 is a loan guarantee program similar in effect to a debt moratorium. But, the requirement that creditors must write off 10 percent of the debt in order to obtain the guarantee introduces a new angle. On the surface the requirement appears to make the lender share in the loss. However, the situation changes when viewed from the lenders' perspective. Lenders are not likely to write off principal where the debtor has a reasonable chance of success. On the other hand, many farmers are already technically insolvent. The lender has substantial incentive to write off 10 percent of an already doubtful debt in order to insure the remaining 90 percent. From this standpoint, the loan guarantee program is a subsidy to agricultural lenders who participate in the program. For the technically insolvent farmer, the effect of it resembles that of debt moratoria except that, at this point, he has little left to lose.

144

Alternative 4: <u>Restrictions on individual rights to contract for mortgage debt on land could be utilized as a measure against over-indebtedness</u>.

Since the basic financial problem we are addressing does involve people overinvesting at excessive prices, it can be argued that we could legally restrict that kind of activity. Several means to effect such a policy come to mind.

For one, legislation might allow a land buyer who finds himself in trouble to default on certain kinds of land contracts. Most find that response so unattractive they never even think of it, but consider, though, that at least some of the sales made close to the peak of the land boom required seller financing through second mortgages. In these cases the seller may have made substantial windfall gains on his holdings and surely recognized the risk he took when he agreed to the second mortgage. As a result, this person may be the socially preferred choice to stand the loss which inevitably follows from a decline in farm real estate values. If that is true, policy could include encouraging troubled buyers to default.

Pursuing this idea further, it might be useful to place restrictions on future land contracts and mortgages by requiring them to allow default without attachment of other real or personal property. Such a provision should lower the price of land during a boom period and reduce the opportunities of people with limited cash resources. It might also increase interest rates for mortgages. But in fact, many of the individuals who are now suffering severe financial stress were those who had limited resources, limited at least for operating on the scale which they chose and which their managerial skills suggested was realistic. In effect, the right to speculate on land values through the use of financial leverage might be restricted by limiting contractual possibilities. So that kind of policy would moderate future land booms and, so moderate busts as well. Few, probably, object to the idea of moderating busts. But few are likely to embrace the idea of limiting booms.

Policies of this kind have several historical models. Fedwer (1952) points to two examples--one Swiss and one American. During the depression of the 1930s and again after WWII, the Swiss government decided that it needed to limit excesses of the private enterprise system and to constrain certain constitutional freedom. As a result, it enacted elaborate overall measures against overindebtedness.

The Swiss passed legislation in 1940 for the purpose of both reducing current debts and preventing excessive indebtedness from occuring in the future. The Swiss bill restricted "the liberty of farmers to get themselves into debt" (Fedwer, p. 238). This law limited encumbrances to the extent of the value established by the appraisal of the land, at the owner's cost, by a state appointed appraiser. It changed inheritance laws so that economic units sufficient to support reasonable operation had to transfer in their entirety to a single capable heir at a price based on the normal income earning capacity of the unit in question. Further, farmers who acquired such land could not sell these farms whole or in parcels within six years following the purchase--presumably to limit arbitrage over time, which is to say short-term speculation, in land.

A final aspect of the Swiss bill covered the liquidation of debt when it exceeded the appraisal value of the farmer's real estate. The objective of this provision was permanent consolidation of the enterprise and adjustment of debt to a normal level. The provision was to be applied only in specific cases where the farm constituted the material means of support of the farm family, would not cash flow under existing debt, and the farmer's distress was not the result of his own poor management. Liquidation broke the relationship between debtor and creditor and created two new contracts: one between the debtor and the state and one between the state and the creditor. Creditors received a combination of dividends based upon the farm's earnings and state guaranteed bonds. The higher the debt, the higher proportion of claims which were satisfied by dividends. The state obtained a non-interest bearing mortgage on the property. This in effect regularized and mitigated what might happen in an unfettered market:

> Thus the liquidation is a forced amortization or scaledown; a subsidy in as much as the nation participates in reducing the creditors' losses; an intermediate solution because the creditors obtain government bonds in place of an uncertain claim against the distressed farmer, and the latter stays on his farm; and it is limited in time and scope (Fedwer, 1953, p. 238).

The Frazier-Lemke Act, Sec. 75 of U.S. Bankruptcy Laws entitled "Agricultural Compositions and Extensions" which expired in 1949 had a similar purpose. Under Subsections (a)-(r), passed in 1933, a farmer could propose to his creditors either an "extension" or a "composition" or both. An extension is an agreement whereby the debt repayment schedule is extended. A composition is an agreement for the payment of a percentage of the creditor's claims in full while the farmer is permitted to retain the farm if he makes the payments agreed upon. In each case, a majority of the creditors, representing more than half of the total amount of claims, had to accept a proposal for the composition or extension.

Under Subsection (a), added in 1934 and as amended in 1935, a farmer could ask to obtain a three year moratorium, during which time he maintained possession of his farm under the supervision of the court, if he paid a semi-annual rental. After three years, the law gave him the right to "buy" the farm by paying to the creditors a sum equivalent to the farm's "then fair and equitable market value." Court appointed appraisers determined market value.

Subsequent judicial decisions, especially by the U.S. Supreme Court, later filled the many loopholes in this emergency legislation. Also, many policy makers proposed incorporating all of Section 75 into a permanent Chapter 16 of the Bankruptcy Laws. Others argued strongly that the current legislation, particularly Chapter 12 of the Bankruptcy Laws on "Real Property Arrangements by Persons Other Than Corporations" should amply protect farmers. And Chapter 12 allows for the proposal of a composition. Ultimately, U.S. policy makers rejected this policy approach, for philosophical as much as for practical reasons.

So, as consequences of restricting contractual rights with regard to mortgage debt, we might expect

- this policy to moderate subsequent boom-bust sequences and to stabilize, or render less flexible, farm ownership patterns
- such a policy to restrict opportunities for people with limited cash resources
- over the long run, to restrict government expenditures for research and development concerning more efficient forms of economic organization.

147

Alternative 5: <u>A supply of government land could be maintained as a buffer stock</u>.

Although we recognize that it is extremely unlikely that the government would be able to efficiently manage the purchase and sale of land in order to stabilize prices, the concept is interesting because of the potential for a ready supply of set-aside acres. Further, if erosive land was purchased, the federal government could perhaps make a socially useful determination to "save" that land for later use.

The cost of such a policy might not be totally out of the question. Eight billion dollars each year would currently buy about 1 percent of U.S. land annually. Roughly 2 percent of U.S. farmland changes hands in an average year, so an amount similar to the cost of recent agricultural programs might very well stabilize the price of land.

This program could begin with land acquired by the Farmers Home Administration through defaults and foreclosures. The government could rent this land back to qualified farmers, perhaps even the original owners, and they could be given a repurchase option. Harl's proposal (see below) for interim land ownership could supply the mechanical details.

Many of the contractual arrangements of such a program could be copied from the several private attempts to purchase land from financially troubled farmers. While organizers of these private attempts are now having trouble obtaining public approval from the states they wish to operate in, forming several nationally sponsored (and possibly regulated) private corporations could obviate that problem. Stock ownership in these corporations could be widespread with the result that land ownership would be comparably broad-based. Formation of several such corporations could foster competition for the efficient management of the land resources.

So, as consequences of a policy for public purchase of land--whether through a government agency or a regulated private corporation, we might expect

- the price of land to stabilize
- promotion of more widespread ownership of land resources
- the nation to gain the ability to manage more erosive lands directly and to acquire a ready source of set-aside acreage.

148

Alternative 6: <u>A government funded program for counseling individuals about their specific problems</u>.

Part of the pain which we desire to alleviate is probably one step removed from the actual financial losses. Perhaps much of the pain derives from the act of solving the problems associated with financial losses--the pain of thinking, analyzing, and communicating with the lender.

Committees of local farmers and qualified farm managers were to be given responsibility for working with farmer-borrowers and lenders to develop financial plans, farm by farm, in the farm credit initiative announced by President Reagan on September 18, 1984. Paarlberg has discussed the strengths and weaknesses of such a program, and observed that these committees are at the heart of the farm credit plan for debt restructuring. He stresses the need for tough-minded and discerning review of individual plans.

When a farmer has devoted a lifetime of education and work to developing a set of attitudes and to creating what conventional wisdom deems an optimal farm operation, then there is a formidable psychological effort required to overturn that thinking, to recognize that all is not what it seems, that conventional wisdom was not necessarily so wise, and that the nature of the operation must undergo change. All of that cuts close to raw ego. It hurts--acutely.

A lucky few might find help and comfort from friends who can aid and support them through all of this painful thinking and negotiating. Society could institutionalize all of this through the use of committees composed of neighboring farmers and trained financial analysts--people with the expertise and knowledge of local conditions to be able to offer substantial help. During the Great Depression, this kind of mutual support system was widely used.

Alternative 7: <u>Interim land ownership and financing</u>.

Neil Harl has proposed a federally chartered entity to buy farm land at current market prices from financially stressed farmers, a device which could provide more short-term aid during the financial crisis than could debt moratoria and loan guarantees. This approach injects outside equity, either government or private sector funds, into the farm sector to cover existing debt. By design, it adds liquidity in the real estate market and provides some support for farmland prices.

Harl's entity would take title to the land and assume an amount of indebtedness equal to the current market price. The creditor would, then write off any additional debt previously secured by the land. In most cases, the entity would rent the land back to the previous owner at market rental rates. Interest payments in excess of the revenue from rent would be made up by the owners of the entity (government or private). The original debtor owner would have the opportunity to repurchase the land at market values at a specified time. Under Harl's outline, the arrangement would be temporary and the intention would be to resell all land within four or five years.

Harl calculates that a congressional appropriation of roughly $2.5 billion a year would generate enough financing to make interest payments in excess of rental revenue. To the extent that the initial market values at which the land would be acquired represented the discounted value of the rental stream plus anticipated future capital gains, the government would recover the $2.5 billion annual contribution (with interest) at the time of the sale back to the original owners. However, we find a contradiction between the assumed effect of initial support for land prices and the implied assumption that prices would not be subsequently depressed when the entity resells the land in four years. Perhaps the resale of land would need to be spread over a longer time.

Alternative 8: <u>A futures market for land would allow the transfer of risk</u>.

The major factor causing harm in land booms is the speculative element of asset ownership. We use the imprecise phrase "speculative element" here to refer to our inability to know what changes in asset values will occur in the future and the implied risk associated with holding such assets. Financial crisis situations result whenever general beliefs about future returns to farm assets are substantially revised downward. The value of farm assets falls by an amount which was previously believed to be highly unlikely (though perhaps possible) and asset holders suffer quite large losses. The condition of farm structure which makes it so easily susceptible to this is the difficulty of participating in the business of farming while avoiding speculation on land. A number of factors summarized earlier in this chapter result in an economic structure where at any point in time many individuals will not survive a severe and unexpected decline in farm asset values. That economic structure results because the costs of avoiding this risk are currently quite high. A futures market for land could lower that cost.

150

Forget, for a moment, concerns about the impractica-
bility of such a market. Developed conceptually, such a
market appears quite feasible. Operational details are
discussed in the final chapter. Please consider here the
social value of solving the basic problem and the theoretical usefulness of a futures market for land.

A futures market for land could allow the divorcing
of the speculative element from the remaining attributes
of land ownership and stewardship. Farmers wishing to
purchase land, with or without limited resources, would
have the opportunity to shift the risk of changes in land
value to others better prepared to stand that risk. The
land value could be "hedged". The property's value might
be hedged on the farmer's own volition or at the insis-
tence of the lender.

It is important to recall, in evaluating the benefit
of a futures market in land, that price changes in the
futures market would be directly related to the expected
income streams attributed to these resources. In developing our conception of the farm problem we began with a
discussion of profitability of agriculture and cash flows
from an accounting framework. We argued that land price
adjusted to reflect expectations for these flows in the
future. By the same reasoning, fluctuations in the value
of the futures contracts on land would tend to offset
declines (or increases) in farm income. This, in turn,
would offset cash shortfalls due to the purchase of "over"
priced land or cash overages due to the purchase of
"under" priced land.

The function of a futures market for land--transfer
of risk--has been traditionally handled through leasing.
A significant, though variable portion of the nation's
farm property has long been owned by absentee landlords.
These individuals lease the rights to utilize the land for
production to farm operators. Although leasing may function to transfer risk of land price changes more perfectly
than a futures market could,[1] leasing involves more
responsibility than an absentee owner may care to shoul-
der. Further, leasing leaves any tax implications associated with land ownership with the landlord. During the
boom, the latter attribute of land ownership encouraged
substantial outside investment interest.

This alternative is more fully developed in the next
chapter. Consequences of organizing a viable futures market for land would include:

151

- The major speculative element of land ownership, i.e., changing land values, could be separated from the primary business of farming.
- The level of equity required by a farmer owning his own land would be greatly reduced.
- For a given level of equity a farmer could greatly increase the probability of his business survival.
- Individuals throughout the nation, or world, having information regarding the general level of future land prices could express their opinions through land investments without the additional trouble and cost of managing specific properties.
- Government expenditure required to implement this potentially fruitful alternative would be restricted to research and development of a viable futures market for land.
- Development of an institution for organized trading in futures on land represents one of the few policy options with potential to ameliorate the basic problem in the long-run.

Summary

Price and income support policies cannot address the root of the problem as we conceive it. Rather, attention must be directed toward institutional changes which make our system of farm asset ownership more resilient to the fact of imperfect income forecasts. We suggest possible broad approaches to this end.

Restricting individual rights to contract mortgage debt on land could be utilized to discourage over-indebtedness. This approach could moderate the effects of boom-bust cycles. However, the current limited opportunities for individuals with low net worths would be further reduced. Minimum equity requirements of $250,000 to $750,000 to maintain a viable sized family farm with high probability of survival are already high and would be increased under this approach. Such equity requirements might be reduced under another economic organizational approach. This option is not in the spirit of our free enterprise system, and we would not endorse it.

One practical alternative is directed toward dealing immediately with thousands of individual financial problems. We stressed working these problems out on an individual basis and emphasized the advantage of utilizing committees made up of local farmers and farm managers.

Our final alternative provides for a long-term solution to the farm problems. A futures market for land would allow the transfer of risk to individuals or groups better endowed for sustaining that risk. This is accomplished under our present system through absentee land ownership. However, this system requires the absentee owner to assume much more responsibility for management of the land than would be necessary under our proposed alternative. A related alternative involving acquiring and maintaining a government owned stock of land might be utilized to soften the current adjustment and could serve as the fungible base for a futures market in land.

Our proposed alternatives represent fundamental changes in the traditional approach to farm problems. Our view is sufficiently important to bear repeating. Farm income is not a farm problem. In so far as income forecasts cannot be improved upon by government, policy efforts should be directed toward making our system of economic organization better adapted to handling inevitable forecast errors.

NOTES

1. The futures market must be for a general "representative" type of land rather than for specific parcels of land. The value of "representative" land would not change in exact proportion with that for any specific parcel. But, the respective decreases under a substantial downturn would be expected to reasonably proportionate.

7

A Policy Proposal

We have suggested, throughout our discussion, that
asset values and an imbalance between debt and equity
financing lay at the heart of our crisis in agricultural
finance. This apparently counters the thinking of most
designers of farm policy, for all of the policies enacted
since the depression years appear to follow from a very
different notion--that market prices for commodities or
income levels are the source of the problem. To be sure,
prices do not always achieve expectations or needs. And
financially troubled farmers do need more income. But
those seem more symptoms than root problems. Indeed, we
think it highly significant that many of the traditional
policies seem to make things worse for farmers.

When supposed solutions actually create additional
problems, then it is time to look for a new conceptual
framework, to engage in some "zero base" thinking. Once
the proper conceptualization is achieved, solutions often
suggest themselves. For example, once we entertain the
possibility that asset values and the system of farm land
ownership are at the heart of the problem, our thought
turns away from price supports and other such measures.
Then, when we see the problems inherent in our typical
method of financing land purchases, we see that a major
share of the problem stems from the difficulty certain
types of farmers are likely to have because of the volatility of land values.

Farmers, and others, have dealt with market volatility for generations by hedging in futures markets. That
effectively insulates them from certain kinds of risk.
Thus a parallel solution might be an appropriate way to
deal with the asset value problem. A futures market in
land, we suggest, which allows a farmer to hedge his land

purchases could provide an effective way for him to transfer risk to others better able to bear it--and willing to
do so in expectation of earning a profit. Obviously, such
an approach cannot actually deal in specific parcels of
land. Rather, it would depend on the establishment of
some business unit, private or governmental, which could
deal in equity shares of suitable generality. In our
view, such a structure could considerably strengthen the
farm economy and make it more resilient--better able to
cope with both extremes of a boom-bust sequence.

So far in our discussion, it has been enough that the
idea of a futures market in land seem appropriate and useful. But having established the nature of the farm financial problem and having provided a conceptual framework
which highlights the limitations of traditional farm policy, we are ready to show how such a futures market could
be established, how it might operate, and what its effects
would be likely to be. To begin with, though, a brief
overview showing how a hypothetical farmer could benefit
from our proposal will provide a context for the more
technical discussion which follows.

Consider a young, beginning farmer who has sufficient
equity to safely finance operation of a farm, but insufficient equity to finance the purchase of land. Our proposal would provide potential for this person to own land
with minimal risk of failure from unexpected economic conditions. Assume that there exists a corporation which
holds a portfolio of farm real estate. The portfolio is
representative of land such that when the value of farm
land changes because of revised expectations, the value of
stock in the corporation changes accordingly. Further
assume that there exists a market for futures in the corporation's stock.

Now suppose the young farmer wants to purchase
$100,000 worth of farm land. His banker tells him that
his equity is insufficient unless he will hedge the value
of the land through taking an offsetting position in the
futures on stock in the land holding entity. Doing that
relieves the farmer of the risk of declining land values
of the sort which have recently put so many farmers in
severe financial straits. Of course, the farmer would at
the same time forego opportunity to participate in appreciation in the value of his purchased land.

The procedure might operate in the following manner.
As part of arranging debt financing for $100,000 worth of
land, the farmer would sell short futures on $100,000
worth of stock in the land holding entity. The bank would

loan, say, $75,000 for the land purchase and finance the farmer's futures position. Suppose further that the farmer's cash flow situation is such that income on the land covers the cost of debt. Now, if land prices fall by 25 percent, there would be a $25,000 loss on the farmer's equity in land and his equity would have been wiped out had he not also made the futures transaction. However, the price of stock in the land holding entity would also fall by 25 percent, due to the fact that the corporation's value is entirely determined by the value of its land holdings. Therefore, the short position on the stock makes a $25,000 profit. Profit in the futures position would be used to pay down debt on the land by the same amount that the value of that land had decreased leaving the farmer's equity unchanged. The $25,000 profit is immediately used to repay debt. Cash shortfall for the farmer due to a reduction in expected income from the land is offset by the reduced level of debt and interest expense on the land. The reverse holds for increases in land value.

By itself, the idea of a corporation which acts as a holding company for land is not new. Harl's proposal contains such an entity. And, in fact, large companies, such as insurance companies, often own vast tracts of farm land --much as Harl's, or our, corporations would. The unique aspect of our proposal is its futures market for farm assets. As our very informal example suggests, the potential of a futures market in land is worth careful exploration and thought. In our view, this approach has the potential to face the crucial financial problem head on and, as a side benefit, introduce a considerable measure of stability into the land market.

A Futures Market for Farm Assets

The key feature of futures exchanges for commodities is that they facilitate an exchange of the risks that derive from the impossibility of knowing just what future commodity prices will be.[1] Specifically, these exchanges provide for shifting risk which ordinary insurance cannot cover. And we think the risks arising from the uncertain nature of returns to farm assets, and from the changing values of those assets, closely parallel the commodity situation. Farmers, or other land investors, cannot average out these risks over usefully short periods of time

157

because unexpected changes tend in the same direction as the boom-bust sequence progresses. In fact, as more people, and assets, enter the pool, the total risk would increase. Instead, the market must adjust these risks over time, with outcomes subject to forecast error. This process of covering this type of risks is what people ordinarily refer to as speculation.

Coverage of the risk which follows from changes in the returns to, and values of, farm assets in a way which could make farm success less uncertain requires that someone commit vast equity capital. Given that accurate forecasting of those returns and values is impossible in principle, information which contributes to more accurate forecasting would become extremely valuable. In this context, a futures exchange which would allow farmers and investors to easily exchange that risk would have much to offer. In general, a futures market in land would add important flexibility to our land ownership system and make it more resilient in the face of our inherent inability to forecast returns accurately.

More specifically,

- Liquidity in the land market could be enhanced by means of a futures market for land.
- Aggregate forecasts for returns to land could potentially be more accurate and be transmitted throughout the economy in a more efficient and timely manner. Such a market could bring additional information from many diverse sources to bear on the problem of forecasting returns to land ownership. This benefit would result from lessening the transaction costs involved in arbitraging expected returns.
- Opportunity to separate the functions of land stewardship and farm business operation from that of covering the risk due to uncertain returns could increase the efficiency with which each function is carried out. That gain would be due to the potential for specialization.

Some Aspects of a Futures Exchange for Farm Assets

Granted the general attractiveness of the idea of a futures market in land, we need to consider in some detail how such an institution could serve farmers. The farmers most vulnerable to the vagaries of the financial situation

158

which farm land ownership entails, as we have observed, are those who are just starting out and those who are trying to increase the size of their operations to gain economies of size and the greater profitability which that brings. Both groups, given the present structure of the farm economy, tend to carry heavy debt loads--a risky business for even the shrewdest managers. Within our proposal, though, the institutional structure would enable such farmers to neutralize those risks by assuming opposite risks. That is the essence of the hedging strategy.

Consider, for example, the case of a young man who wants to begin farming, as an owner-operator, on a scale such that his farm would be economically viable. The beginner must, from the outset, have equity capital sufficient to weather the regular business risks of farming-- those associated with such matters as variable yields, price fluctuation, and machinery repairs. Of those who qualify in terms of managerial ability to undertake such a problematic venture, few have the good fortune to be wealthy enough to be able, also, to bear all of the risk of changing land values.

Our young man, given the present ownership system, may simply have to forego his ambition. However, if there were a futures market like the one we propose, such a young man might transfer the risk of land ownership by selling futures on his land. The current financial crisis in the farm sector, stemming as it does from declining land values, suggest the possibility that many people in the farm sector might like to take similar steps to neutralize risk. Someone like our young farmer could enter the futures market as a hedger-seller while holding an opposite cash position through actual land ownership.

Of course, sellers alone do not a market make. There must be buyers, too. While we can imagine other farmers, or potential farmers, who might assume the role of hedger-buyers, we doubt there would be enough of these people to balance the market. Ordinarily, among farmers, we would expect hedger-sellers to predominate. As a result, there would be a risk surplus for which opposite hedgers would not exist. Ordinary insurance converts this risk surplus by calculating a reasonable cost for it. While risk pooling insurance is not possible, in the case of land, the concept of a risk premium is useful. The only solution is to transfer the risk surplus to people willing to bear speculative risks for expected profit. In other markets, speculative traders are willing to enter into a futures transaction because they expect their superior foresight to make their risk-bearing profitable--usually not because they are trying to insure against other risks in that market. Similar thinking would motivate the entry of such traders into a futures market in land.

In the general case, speculators would outnumber hedgers, as surveying a week's transactions at, say, the Chicago Board of Trade demonstrates. Some would be buying, some selling. All would be united by their striving to gain through superior foresight, or ability to predict or anticipate changes in land values.

Speculators serve a market in a variety of ways. Besides providing a liquidity which the land market now lacks, the speculators would function valuably in acquiring and synthesizing the information so vital to fruitful judgements in such a field. Bringing speculators into this market should be a positive move. For broadening the number and kinds of participants in the land markets should broaden the kinds of information available to all the participants. That should, in turn, allow all participants in the land market to increase their insights and sharpen their decision making.

As long as there were an excess of hedger-sellers, relative to hedger-buyers, the market in land futures would need to generate and nurture more speculator buyers. One simple way to do that would be to offer speculators a risk premium. This device, which Keynes called "normal backwardization", would cause futures contracts to trade at an actual price slightly below the expected nominal future cash price.[2] That is, a hedger-seller, like the young farmer buying land, would place his hedge fully expecting to take a loss on the transaction equal to the risk premium.

While a risk premium might attract speculators, the futures market, to be successful, would have to take further steps to ensure the continued satisfaction of this group. In any market, speculators want to filter out all risks not related to the primary issue. That is, speculators in farm land futures would want to limit their risks to those directly related to changes in land prices. For instance, they would insist upon security with regard to fulfilling contract obligations. When a market manages to exclude all risks except those relating to price, we say we have achieved "perfection of the market". That condition pleases speculators, but it is also essential to hedgers.

160

Futures markets strive for market perfection and liquidity by standardizing the quantity and quality of the commodity they trade, the qualifications of the traders, and the trading methods. There is often a tradeoff between the benefits of rigidly standardizing the commodity and the dangers of creating natural or artificial scarcity of the standard grade. Some markets create a degree of flexibility by allowing the seller limited discretion concerning commodity grade and delivery time. With regard to futures in farm land, this is a challenging issue. It is not at all easy to see how to design a standard for farm assets.

It is also important to consider how buyers or sellers of futures in farm assets could demand delivery of the commodity--or enforce the acceptance of it. Clearly, divergent movements of cash and future farm asset values would impair the effectiveness of hedging in direct proportion to the degree of divergence. Yet, rights concerning delivery of the commodity can govern parallel movements of cash and futures values.

Any market in land futures must overcome these obstacles if it is to function in a useful way.

As we move toward a definition of how a futures market could work, we need to keep in mind several important characteristics which the exchange media will have to exhibit if the market is to operate effectively:

- The contracts should capture changes in long run returns to farm assets, or equivalently the value of farm assets.
- A market determined risk premium could be reflected in the market as backwardization of prices, as an up-front payment or as an annuity paid by the seller to the buyer.
- The market should contain a minimum of risk in excess of that due to changing asset values.
- Contracts would have to be effectively deliverable in order to force the parallel movement between spot prices on the exchange and cash events in farm asset markets.
- Quality, quantity, and method of trading would need to be standardized with perhaps some degree of flexibility with respect to grade.

161

One Approach to Trading in the
Future Values of Farm Assets

We could usefully shift the risk which arises from changes in asset values in either of two quite different ways. In a sense, we can view the purchase of land as the acquisition of rights to an income stream. So a futures market in land could either trade contracts on actual assets, or it could trade equivalent rights to future income streams. While either would achieve the desired ends, we base our proposal on the trading of contracts on actual assets. It seems simpler and more satisfying.

To establish the market we envision, there would first have to be one or more business entities, capitalized through the sale of equity shares, for the purpose of purchasing and maintaining a representative portfolio of farmland. The organization and management of these entities would require careful attention in order that the value of their stock would parallel farm asset values. That attended to, though, the operators of the portfolios could institute a futures market for equity shares or stocks through which farmers could neutralize the financial risks related to falling asset values and which would entice speculators to cover those risks through the attraction of market determined risk premiums.

The Commodity--Stock in a Portfolio of Farm Land

To be tradeable, a commodity must lend itself to standardization and deliverability. Off-hand, farm land seems not to qualify on either count. We do grade crop land according to productive capacity, even that involves many important variables which often require subjective evaluation. So grading land appears unworkable in the context of this kind of market.

Yet, if the unit of trade were a share of stock in the portfolio of representative land holdings of the corporate entity, we can see how useful trading could go on. But even then the portfolio would need to be large enough to cover the demand of short selling by hedgers and to reduce the possibility that anyone could manipulate the stock. Finally, it might best serve interests on all sides if there were separate portfolios to represent types of farms--such as midwest corn-soybean, western wheat, or southern corn-soybean farms. Such a set of portfolios would be representative of the farm sector nationally. That set could then serve farmers generally since everyone could find a trade medium appropriate to his needs.

Management

An important factor in the success of the futures market would be the management approach of the firm holding the portfolio of stock in farm land. For the market to attain the highest possible degree of perfection, returns for the portfolio assets would have to closely parallel returns for that farm type. And the relationship between portfolio returns and returns of that type of farm would have to be relatively stable. Because there could be little tolerance for management error, one useful management strategy would be to assign partial responsibility to each of several management groups. We could reasonably expect that, in such a case, income variations resulting from management would average out within each portfolio in the set.

Operational Detail of the Risk Transfer

Hedging, essentially, offers a farmer a practical means of distributing risk. We cannot make the risk go away, so we look for a way to distribute it among people better able to bear it. A futures market facilitates that transfer of risk. That is the essence of hedging--to allow the farmer to minimize the degree of uncertainty he must contend with as regards changing asset values. Such a futures market for land also moves us closer to the Jeffersonian ideal of widespread asset ownership in a very real way. For when land assets can be freely exchanged by both farmers and nonfarmers. Then people enter into these transactions because they want to. That kind of free participation enhances the shareholder's guise of the importance of land ownership.

A corollary gain--the farmer also reduces to some extent the risk associated with debt financing. In fact, the futures market would generally reduce the degree to which the sector depends on debt financing. Upon buying land for his own operation, a farmer, as hedger-seller, could hedge his risk by selling short an appropriate amount of stock in the company or agency which maintains

163

the portfolio of land. He would have to expect to pay a market determined fee for the privilege--in effect, an insurance premium.

To offset such short positions, the futures exchange would offer long positions to speculator traders, and would entice the speculators to do so by offering them the same market determined risk premium.

Hedger and speculator, alike, would have to maintain some amount of margin deposit, to cover potential losses. For speculative positions, existing futures exchanges provide a viable model. But hedger-sellers might benefit from a different arrangement. As simple and practical as anything would be a structure which allows farmers who are hedging to obligate a portion of the annual farm income on the land.

A hedger-seller would neutralize his risk to the extent that the value of the stock in the land portfolio rises or falls in concert with the value of his own land. When land values fall, as they have recently, equity in a proportionate short position would rise by a like amount. Since falling land values result from declining income, and negative expectations concerning income, a farmer would experience a restricted cash flow at the same time. Given that, there would have to be provisions made for him to draw upon his equity in the stock position to reduce his mortgage debt. The cash that he requires could be available as a result of the requirement that speculators put up additional margin monies to replenish their own declining equity positions. The reduction in mortgage debt reduces the farmers interest expense and thereby restores his cashflow position.

Initial Capitalization and Start-up
of the Land Holding Equity

Initiating this kind of financial activity requires vast funding and careful management. For not only would the portfolio have to be extensive to provide the desired results. It would also be necessary for the land holding organization to strive for profitability. Only then could it function effectively in the economy. Also, the management approach has to serve other ends than making a profit. For one, this financial structure needs to provide aid to certain troubled farmers. If it fails to do that, it fails in spite of its profits record. It needs to introduce a measure of stability in the entire land market operation. Fortunately, there seem to be a variety of approaches which could accomplish all of that.

One approach to capitalization and start-up is that suggested by Harl (1985). To form his "Agricultural Credit Corporation" (ACC) for "interim land holding and financing", Harl proposed purchasing land from financially strapped farmers and assuming responsibility for an amount of debt equal to the current market value of the land. The ACC would receive revenue from current income on the land and use it to pay the interest cost on the assumed debt.

In general, then, Harl's proposal seems to provide a useful device for getting under way. However, the effect which using financial leverage, as Harl contemplated doing, would have on the behavior of the stock prices of the land holding concern could create serious problems. Obviously, for such an institution to perform its intended work satisfactorily, stock prices must clearly reflect changing land values. Using financial leverage would no doubt interfere with that relationship unless there were an appropriate control arrangement in place.

In our view, Harl's approach uses too much debt. We recognize that in the first stages of a new venture it may be hard to sell stock. So a structure like Harl's might provide a useful transition. Basically, though, we want to let stock sales carry the program as soon as possible. That makes the structure we propose far simpler and more direct--all in all a more reliable reflector of the land market and so a better tool.

One possible way to avoid leverage, and to improve on Harl's approach, would be for the federal government to assume responsibility for the initial purchases and for the debt quite apart from the land holding firm. Using revenue from its stock issues, the land holding entity could buy out the land which the government purchased and then allow the government to retire its debt at that point. By regulation, financial leverage would not be an available strategy for the land holding entity. Alternatively, the federal government might underwrite the start-up phase. The ultimate aim in any case would be to have funds from the land holding firm's stock issues immediately represent the market value of the land holder's land assets on a one-to-one basis.

All of the provisions for capitalization and start-up which we have discussed serve the long-term financial needs of the farm sector. Important as that is, the short-term needs of the sector must also receive

attention. Actually, providing short-term aid is the main
reason to try to speed the start-up. A rapid transition
would provide a bridge over which those currently in distress could establish a foothold which would permit them
to benefit from the long range possibilities of the system
we propose.

A farmer who cannot finance his current situation
cannot take advantage of an opportunity to hedge against
further declines in asset values--especially if there is
the additional cost of the risk premiums. But the initial
phase of the program we envision provides for a significant easing of the current situation. Our policy proposal
would lead to an improvement in the liquidity of the land
market. It would create an opportunity for farmers in
various categories of financial stress to sell land in a
relatively advantageous way. In short, there are broad
opportunities for immediate help for many who cannot now
see any forthcoming--and all without substantial governmental budget outlay because the obvious initial source of
land for the holding entity would be farmers who face
immediate foreclosure. To the extent that the land holding company reduces transaction costs and establishes
trust in its stocks, it will enhance the demand for farm
land as an investment. In turn, that would reduce the
extent to which land values fall. And any stabilization
of land prices would ease the situation of those now facing financial difficulty.

The organizational structure of the land holding
concern, further, could be such that financially stressed
farmers who sell land to the holding company could receive
preferential rights concerning repurchase of the land and,
in appropriate cases, continue to operate as tenants. The
repurchase could make use of call options--at a predeter-
mined price which would give the holding company a reasonable profit. Another way to handle repurchase would be to
charge the then current market price. In that case, the
original owner could hedge against the risk of price
changes by using futures contracts on the holding firm's
stocks. A third way to handle it would be simply to issue
a first right of refusal to the original owner.

A Numerical Example of Hedging
Against Declining Land Values

While we cannot gain the benefit of watching such an institution actually in operation as we evaluate this proposal, we can evaluate it in a reasonably useful way through consideration of a hypothetical example which we base on some fairly conservative numerical suppositions. To begin with, suppose that the land holding entity (LHE) purchases 3 million acres of productive land at market value. Expected return to the land will be a primary factor in determining market values and guide the land holding concern in its purchases. Suppose that, on average, we can expect the land to return $72 net per acre in the first year and that the amount will increase with the general rate of inflation. Assuming 4 percent anticipated inflation and a 10 percent nominal discount rate implies that the land should sell for $1200. The total investment would then be $3.6 billion dollars. Current income from the land would be expected to average 6 percent of the total investment.

If the 6% real rate of return is acceptable to investors, the land holding entity could issue 3.6 billion shares of stock at $1 a share and purchase the 3 million acres with the proceeds. Then, trading in the LHE stock could begin. As trading in the stock futures began, the hypothetical 4 percent rate of inflation would cause a $1 share of land stock to be worth $1.04 after a year.

The land holding concern would pay annual dividends from the revenue that the net operating returns to land would generate. In our example, we would expect those dividends to average 6 percent of the stock value. Also, we would expect the market value of the stock to vary directly with the market value of the land. That should follow from the fact that both reflect the present value of future land earnings. In the absence of expectations of change in relative prices, land and stock should both appreciate at the general rate of inflation--about 4 percent per year.

During a so-called normal period of stability before a boom, that probably describes the operation of the land market. But, when events occur which alter our beliefs about future income returns or inflation rates, we might expect rather different behavior in the market. In general, land stock value will change so that investors can maintain the acceptable 6 percent real rate of return on investment.

Suppose, for example, that for some reason we expect real future returns to average $60 an acre, a 16.67 percent decrease from the $72 we had initially expected. At the same time, we expect inflation to maintain a steady rate, and we continue to require a 6 percent real rate of return. Given all of that, the value of land should fall to $1000 an acre, or 83 percent of the initial value, and the value of land stock should fall a like amount, to $.83 a share. Expectations of increased real returns motivate a parallel, though converse, case.

When a farmer wants to hedge against the possibility of unexpected declines in future earnings, he can insure against risk in a fairly simple way. Given 4 percent expected inflation a share initially worth $1 would be expected to be worth $1.04 at the end of the first year. Anticipating that transaction costs and a risk premium would manifest themselves in the form of backwardization of prices, futures contracts obligating delivery of one share of stock at the end of the first year would initially trade at $1.03. The discount of $.01 represents the market determined risk premium which helps to attract speculative buyers of the futures.

Suppose this farmer wants to transfer risk on $100,000 of land at current values. He can sell an equivalent amount of one year futures in LHE stock. He must obligate equity in his land as security for his side of the position. Because of the risk premium which he has to pay in order to entice a speculative buyer to accept the risk, the farmer can only obtain $103,000 (in our example) for the future stock. So, he expects to pay a cost equal to $1,000 in return for one year coverage on the risk of change in land value. Only three things can happen. Beliefs can remain the same, in real terms. There can be expectation of rising real returns. Or there can be expectations of falling real returns. A potential hedger would want to know what to anticipate in each case.

As we have already seen, if beliefs about returns do not change during the first year of the market's operation, the nominal value of the land and land stock will

168

increase by 4 percent. The farmer's equity in the position will have declined by $.01 per share since the sale was made at $1.03 and the purchase price will be $1.04, so the farmer must deposit $1,000 (100,000 shares @ $.01) with the exchange authority at year end. This $1,000 was the expected cost, and the farmer can view it as an insurance premium which he is willing to pay in order to avoid risk. If nothing changes, this event will occur each year, and each year the farmer will need to roll over his futures position in order to continue the hedge. The first column of Table 7.1 illustrates the result of hedging under these circumstances. The end result is that the farmer's nominal net worth increases by $3,000, which is less than the effect of inflation on the land would have been had the farmer not hedged.

The outcome will be the same even if beliefs about future returns change; however the details on the way to the bottom line will be different. Suppose that we have reason to expect the real level of future returns to be $84 per acre, 16.67 percent higher than the $72 we previously expected. Also, suppose that we expect no real growth in farm returns are expected beyond the new level. Further growth in returns will hold to the 4 percent general rate of inflation. In this case, the price of the land stock will rise immediately to $1.167 from $1.00. In addition to that, the four percent inflation would cause the stock to sell for $1.213 at the end of the year. A one-cent risk premium would mean that futures contracts would immediately begin trading at $1.203, an increase of about $.17 cents a share compared with the no change scenario. The nominal value of the farmer's land will increase by $21,333. But, he will have an offsetting loss if he has hedged through the selling of futures. In fact, the hedging loss will be $.183 a share--the $.173 unexpected price change plus the $.01 anticipated risk premium --or $18,333. Still his nominal net worth increases $3000, as the second column of Table 7.1 shows.

The real value of hedging against a decline becomes apparent when land prices fall because we revise our beliefs about the level of future returns in a downward direction. Suppose events occur which cause us to believe that the long-run level of returns will be 16.67 percent less than we had previously anticipated. Then, the nominal value of the farmer's assets will decline to $86,666 by the end of the year--a nominal loss of $13,333. In the absence of a futures market, our hypothetical farmer would find this a grievous turn of events.

Table 7.1. Possible Results from Hedging $100,000 of Farm
Land.

| | Beliefs About Long-Run Level of Returns | | |
	No Change	Up 16-2/3 Percent	Down 16-2/3 Percent
Ending Asset Value	104,000	121,333	86,666
Beginning Asset Value	100,000	100,000	100,000
Net Change	+ 4,000	+21,333	-13,333
Sale of Land Futures	103,000	103,000	103,000
Purchase of Land Futures	104,000	121,333	86,666
Hedging Profit/Loss	-1,000	-18,333	+16,333
Resultant Change in Nominal Net Worth	+ 3,000	+ 3,000	+ 3,000

Assumptions: 4 percent General Inflation
10 percent Nominal Discount Rate
6 percent Income Return to Land
1 cent per share Risk Premium in LHE 1 year
futures

Given that inflation and other factors hold steady so only asset values change, the hedging farmer will be able to remain solvent and profitable. Where in the other two cases he records a loss from his hedging activity, which partially offsets the positive asset appreciation, the farmer in this case earns a $16,333 profit from his hedging. That more than offsets the loss from declining asset values. The hedger finishes the year with a positive $3000 change in his nominal net worth. That modest gain contrasts sharply with the $13,666 loss a non-hedger would suffer.

These examples show that, whatever our expectations concerning the level of future returns, the hedger operates in a relatively stable atmosphere. And even though his hedging prevents his sharing in the benefits of a boom it also allows him to come through a bust on the positive side. On balance, many farmers, we expect, could see the advantage of hedging.

A Word of Caution

In actual practice, hedging will not produce the ideal results that it does in our hypothetical case. Frequently, returns on a specific farm will not exactly mirror the behavior of the holding company's portfolio. Local conditions will exert their influences in various ways. Returns differ from farm to farm in a given year, and they vary from year to year. No hedging program can expect to smooth out these year to year, or farm to farm, fluctuations.

Yet the futures market should serve admirably in helping farmers cope with conditions like the ones which are driving the current farm financial crisis. In short, the possibility of hedging will allow farmers with high debt to shift the risk of substantial and broad declines in the values of farm assets to other people, in and out of the farm sector, who can better afford to stand that risk.

Implications for Structure

So closely interrelated are the various sectors of our economy that a policy designed to alter the affairs of one sector will surely carry implications, both good and bad, for other parts of the economy. Because of that, we

171

can grade policy proposals according to how many side effects they create, or according to the policy framer's ability to anticipate and control those side effects.

We have indicated that traditional price and income support programs seem to us blunt instruments. Any policy which, on the way to providing somewhat more stable income, increases the use of debt financing in the sector should be viewed with great skepticism. It is certainly not an especially refined policy approach. Also, commodity programs have had strong welfare implications for U.S. consumers and taxpayers and have influenced markets throughout the world. Finally, since these programs encourage expansion, crop programs, and financial practices which market stimuli--or what market stimuli would be in the absence of the policy influence--do not justify, they heavily influence the structure of the farm economy and even farm structure itself. In fact, a little thought suggests that these subordinate effects far outweigh the intended effects of many of these policies. And we agree with many other observers of the American farm economy that the legacy of traditional price and income support programs is not an especially happy one.

In contrast, a policy which strives to increase market efficiency, rather than to control market outcomes, should produce fewer unwanted side effects. Given that, our proposal of a land holding company and organized trading in land futures offers a far more refined approach which actually deals with the basic farm problem while, we expect, influencing other structural aspects of the farm economy to an insignificant degree. Accordingly, the guidelines people have used in evaluating traditional farm policies may largely miss the point.

For instance, our proposal does not impinge on the operation of the agricultural market system at all.

However, this policy approach does promise to effect marked changes in those parts of the economic structure which relate directly to the farm finance problem. As sources in the rest of the economy commit equity to the farm sector, farm financial structure will respond significantly. This will, in turn, reduce the difficulties that come from the concentration of farm sector debt among certain kinds of farmers. This policy will alter patterns of land ownership and of the control of farm operations. It would reduce barriers to entry into the sector which now exist.

Overall, our proposal would make the organization of the farm economy more efficient. Welfare implications would for the most part focus on the improvement of the well-being of farm producers. But, there would also be beneficiaries outside the farm sector. Anyone would have opportunity to invest in farm land. For example, a factory worker with modest savings could invest in farm real estate where now high transaction costs act as a strong deterrent. Still, the basic effect of our proposal on the structure of agriculture will concentrate on aspects of ownership, control and finance of assets in the farm sector (summarized in Table 7.2).

Ownership

Our policy would provide greater opportunity for people with modest net worth to own land. Many farmers currently renting land for cash or on a share basis would like to become owner-operators if they could somehow transfer the risk of broad declines in land values which would quickly erode their equities. The proposed futures instruments could accomplish that. As a result, we expect the share of land farmed by owner-operators to increase given the introduction of our policy suggestions. In addition, land ownership would become a feasible investment alternative for the savings of many people not currently farming. For these people, the land holding company would function much as mutual funds do. Most investors lack enough money, and expertise, to do well in the stock market on their own. Mutual funds allow these people to participate at a level they can handle. Similarly, few people can even hope to buy a farm. But the holding company allows dozens and hundreds of people to join together, in a sense, to enjoy the benefits of farm land ownership. Not only does this achieve the important American goal of widespread ownership of the nations land resources, it does so in a way that allows farms to maintain economy of scale.[3]

Control of Operations

We expect that tenants would farm the land of the land holding concerns. However, we would expect the

Table 7.2. Ownership, Control, and Finance of Farm Sector
Assets.

	Change w/LHE Option
Ownership (Title)	
Owner operators	Increase
Bearing all risk of change in land value	Decrease
Hedged against sector-wide low returns	Increase
Non-Operators	Increase
Individual	Increase
Farm relatives or heirs	Don't Know
Non-farm investors	Increase
Corporate	Increase
Control of Operations	
Land tenancy	Decrease
Cash rent	Decrease
Crop share	Decrease
Part-owners	Increase
Farm size distribution	Don't Know
Diversification	Don't Know
Enterprise mix	Don't Know
Off-farm income	Don't Know
Finance	
Total equity	Increase
Total debt	Increase
Debt to equity ratio	Decrease
Government debt (FHA)	Decrease
Farm Credit System	Don't Know
Commercial banks	Don't Know
Individuals and others	Don't Know

greater ability of tenants to purchase land while hedging
against broadly declining returns to offset any increase
in tenancy caused by establishing the land holding company. In fact, we would expect an increase in owner-
operators and part-owner operators as a result of the
tendency of this policy to open up the market.

How this policy would effect the distribution of farm
size is not clear to us. One advantage which very small
farms now have is their lower equity requirements. The
ability to shift the risk of broadly lower returns would
reduce that relative advantage and allow more people to
operate larger scale farms. However, this risk will not
be shifted without cost. Farmers must pay a market determined risk premium to do so. Accordingly, an operator who
is not making a profit because of his relatively weak production and management skills will not be likely to benefit. Perhaps the major effect on the farm size issue
would be to give management ability greater importance, by
way of reducing equity requirement restraints, as the
determining factor of individual farm size. Overall,
though, the impact on the farm size distribution should be
small.

We also expect the impact of this policy on diversification and enterprise mix to be small. Less diversification might occur if the market determined premium for
shifting risk of declining land values were substantially
less than that perceived by certain individuals.

Financial Structure

Our policy proposal would provide additional equity
to finance the farm sector in two ways. First, investment
in the land holding entity would provide outside equity
directly. As proposed, the new equity would reduce existing debt in the farm sector dollar for dollar. The second
and primary source of equity would be that provided by
speculative buyers of futures contracts on land holding
entity's stock. Such a speculative buyer would be obligating his own equity to cover potential declines in
future farm asset values. At least some of the sellers of
these futures are expected to be farmers using debt. A
hedger-seller will maintain the same debt. But, that debt
will be offset by the new equity committed by the speculative buyer. In sum, a reduction in total debt of the sector might be expected under the proposed arrangement and

substantial new equity would be committed to financing agriculture in the U.S.

Financial leverage in the sector would be reduced in total and, more importantly, the difficulties due to concentrations of debt in the sector among certain individuals would be ameliorated as those with very high debt would be among the first to match their debt with the new equity.

The structure of agricultural lending would probably not change greatly. Success of the proposal might lessen the need for government financing such as that provided by the Farmers Home Administration. But, in general there are no compelling reasons which suggest major shifts in the sources of credit due to the proposal.

Anticipated Limitations

Success of the proposal would hinge first on establishing viable land holding entities. The portfolios of land would need to be representative of individual areas in the nation or types of farms. Returns of the entity would need to be reflective of returns from land in the sector. Attention would need to be given to those aspects of corporate management which affect stockholder confidence and opinion. For example, it would be important to avoid the image which some cooperative ventures have gained where owners feel they have no substantial input in decision-making and no way to market the stock.

The second determinant of success for the proposal is the development of a viable futures exchange for the land holding entity's stock. Use of this futures market will to some extent depend on volatility. However, the expected risk premiums should draw sufficient speculator buyers. Furthermore, initial low volume will by no means imply poor performance since a land boom-bust period will represent the critical test. One would have expected a futures market to have received plenty of attention during the 1970's and perhaps even in the 1980's as participants search for a new equilibrium in the farm real estate market.

Conclusion

We think it quite clear that a futures market in farm land, whatever the exact details of its structure, could effectively protect the farm sector from the devastation of another boom-bust sequence. Given the pattern these phenomena follow, we think it a good idea to work towards that kind of innovation in our agricultural economy right away so that the necessary structures are in place well before the next crisis.

However, the most important idea here is not the futures market proposal. Rather, we think it vital that all those concerned with farm policy recognize that this discussion redefines the problem which farmers confront during these times of crisis. Some time after the enactment of the 1985 farm bills, one farm state senator urged that critics not be too hasty in their judgements of it. One of our points is that any traditional farm help package is bound to fail in the long run because it follows from a misconception of the nature of the problem.

What is important in our study is that we find ample evidence among the standard statistical and historical information to suggest that we need to approach the farm financial problem from a very different conceptual perspective. Once we shift our focus from price and income supports to expectations concerning asset values and the system of farm ownership and finance, new possibilities suggest themselves.

Our futures market proposal is one such. While we value its relative simplicity, there may be other good possibilities. Most important, though, locating an appropriate conceptual framework, besides helping us to identify the problem, helps us to create a systematic means of evaluating a proposal.

In short, we believe our farm families deserve thoughtful attention from policy makers. Their problems are very real, and we cannot afford to ignore either the people or the problems. Offering, as help, something which we know cannot truly work is no kindness. However, we think it possible, now, to move policy discussion to a point where we can come to grips with the underlying problems. And that accomplished, we can find real solutions.

NOTES

1. This discussion draws from Gerda Blau's article, "Some Aspects of the Theory of Futures Trading," Rev. Econ. Studies, 12:1(1944-45):1-30. Parts of Blau's

discussion are paraphrased in the context of farm asset returns and values.

2. Hicks, J.R., _Value and Capital_, Ch. X, pp. 138.

3. Recall that a serious problem of present owner-ship comes to light when a farm, which has become large enough to be economically feasible, must pass to several heirs. Either each must struggle with a unit too small for viable operation, or one heir must buy out the others.

References

AGRIFAX: Comparative Farm Business Analysis, Federal Intermediate Credit Bank of Louisville, Louisville, KY, various annual reports.

Ahsan, Syed M., A.G. Ali, and N. John Kurian, "Toward a Theory of Agricultural Insurance," Amer. J. Agr. Econ., 64:3(August 1982):520-29.

Baird, Frieda, "Reorganizing Our Agricultural Credit Facilities," J. Farm Econ. 15:2(April 1933):319.

Batte, Marvin T. and Steven T. Sonka, "Estimates of Before and After Tax Scale Economies for a Sample of Illinois Cash Grain Farms," Paper presented at the Annual Meetings of the American Agricultural Economics Association, Clemson, South Carolina, July 27-29, 1981.

Bean, Louis H., "Inflation and the Price of Land," J. Farm Econ. 20(February 1938):310-24.

Belongia, Michael I., "Agricultural Price Supports and Cost of Production: Comment," Amer. J. Agr. Econ. 65:3(August 1983):621-22.

Benedict, Murray R., "Agriculture as a Commercial Industry Comparable to Other Branches of the Economy," J. Farm Econ. 24:2(May 1942):476-96.

Black, J.D., R.H. Allen, and O.A. Negaard, "The Scale of Agricultural Production in the United States," Quarterly J. Econ. 54(1939):329.

Blau, Gerda, "Some Aspects of the Theory of Futures Trading," Rev. Econ. Studies 12:1(1944-45):1-30.

Boehlje, Michael and Steven Griffin, "Financial Impacts of Government Support Programs," Amer. J. Agr. Econ. 61(1979):285-96.

Boehlje, Michael and Vernon Eidman, "Financial Stress in Agriculture: Implications for Producers," Amer. J. Agr. Econ. 65:5(December 1983):937-44.

179

Brandt, Karl, "Toward a More Adequate Approach to the Farm Tenure Program," _J. Farm Econ._ 24:1(February 1942):206-25.

Buttel, Fredrick, H. and William L. Flinn, "Sources and Consequences of Agrarian Values in American Society," _Rural Soc._ 40(1975):134-51.

Carter, H.O. and G.W. Dean, "Cost-Size Relationships for Cash Crop Farms in a Highly Commercialized Agriculture," _J. Farm Econ._ 43(1961):264-77.

Center for Agricultural and Rural Development, _Economies of Size Studies_, Proceedings from NCR-113 Workshop on Farm Size, August 3-4, 1983 at Purdue University, Iowa State University, Ames, Iowa, February 1984.

Chambers, R.C., "Relation of Land Income to Land Value," USDA Bulletin 1224, June 1924.

Chambers, Robert G., "Agricultural and Financial Market Interdependence in the Short Run," _Amer. J. Agr. Econ._ 66:1(February 1984):12-24.

Chambers, Robert G. and Utpal Vasavada, "Testing Asset Fixity for U.S. Agriculture," _Amer. J. Agr. Econ._ 65:4(November 1983):761-69.

Chan, Y.L., E.O. Heady, and S.T. Sonka, "Farm Size and Cost Function in Relation to Machinery Technology in North-Central Iowa," CARD Report 66, Iowa State University, Ames, Iowa, 1976.

Dadisman, A.J., J.H. Arnold, and F.H. Branch, "Report of the Committee on Terminology," _J. Farm Econ._ 1(1919):76-77.

Davis, Joseph S., _Agricultural Policy: 1926-1938_. Food Research Institute, Stanford University, California, 1939.

Dobbins, Craig L., _et al._, "The Return to Land Ownership and Land Values: Is There an Economic Relationship?" Station Bulletin No. 311, Department of Agricultural Economics, Agricultural Experiment Station, Purdue University, February 1981.

Doering, Otto C., "Appropriate Technology for U.S. Agriculture: Are Small Farms the Coming Thing? - Introductory Comments," Amer. J. Agr. Econ. 60:5(May 1978):293-94.

Dowell, A.A., "Land Booms and the Mortgage Rate of Interest," J. Farm Econ. 20:1(February 1938):231-32.

Dowrie, Goerge W., "Did Deflation Ruin the Farmer and Would Inflation Save Him?," J. Farm Econ. 7:1(January 1925):67.

Dunn, Daniel J. and Thomas L. Frey, "Discriminant Analysis of Loans for Cash-Grain Farms," Ag. Finance Rev. 36(1976):60-66.

Economic Report of the President, U.S. Government Printing Office, Washington, DC, 1983.

Edwards, Clark, "Modeling Rural Growth," Amer. J. Agr. Econ. 61:5(December 1979):967-72.

Ely, Richard T., "Land Speculation," J. Farm Econ. 2:3(July 1920):121-35.

Englund, Eric, "Fallacies of a Plan to Fix Prices of Farm Products by Government Control of Exportable Surplus," J. Farm Econ. 5:1(July 1923):86-101.

Falconer, J.I., "History of Farm Debt Adjustment Activities," J. Farm Econ. 26:2(April 1934):189.

Featherstone, A.M. and T.G. Baker, "Farm Sector Real Asset Dynamics," Selected paper at the American Agricultural Economics Association annual meetings, Iowa State University, August 1985.

Feder, Ernest, "Farmer-Debtor Relief Legislation in the United States and in Switzerland--A Lesson in Agricultural Policy," J. Farm Econ. 34:2(May 1952):228-41.

Flinn, William L. and Fredrick H. Buttel, "Sociological Aspects of Farm Size: Ideological and Social Consequences of Scale in Agriculture," Amer. J. Agr. Econ. 62:5(December 1980):946-53.

Freidman, Irving, _Inflation a World-Wide Disaster_, new edition. Boston: Houghton Mifflin Co., 1980.

French, B.C., "The Analysis of Productive Efficiency in Agricultural Marketing: Models, Methods, and Progress," _A Survey of Agricultural Economics Literature_, Vol. 1, ed. Lee R. Marvin, pp. 93-206, Minneapolis: University of Minneapolis Press, 1977.

Gabriel, Stephen C. and C.B. Baker, "Concepts of Business and Financial Risk," _Amer. J. Agr. Econ._ 62:3(August 1980):560-64.

Garcia, Philip, Steven T. Sonka, and Man Sik Yoo, "Farm Size, Tenure, and Economic Efficiency in a Sample of Illinois Grain Farms," _Amer. J. Agr. Econ._ 64:1(February 1982):119-23.

Gardner, B.D. and R.D. Pope, "How is Scale and Structure Determined in Agriculture?," _Amer. J. Agr. Econ._ 60(1978):295-302.

Gardner, Bruce, "Efficient Redistribution Through Commodity Markets," _Amer. J. Agr. Econ._ 65:2(May 1983):223-24.

Gardner, Bruce L., _The Governing of Agriculture_. Lawrence, KS: The Regents Press of Kansas, 1981.

Gray, L.C., "Disadvantaged Rural Classes," _J. Farm Econ._ 20:1(February 1938):71-85.

Grimes, W.E., "Social and Economic Aspects of Large-Scale Farming in the Wheat Belt," _J. Farm Econ._ 13:1(January 1931):21-26.

Harl, Neil E., "Public Policy and the Control of Agricultural Production: Discussion," _Amer. J. Agr. Econ._ 60:5(December 1978):844-47.

Harl, Neil E., "Proposal for Interim Land Ownership and Financing," Mimeo, Department of Agricultural Economics, Iowa State University, January 11, 1985.

Harrington, David H., Donn A. Reimund, Kenneth H. Baum, and R. Neal Peterson, U.S. Farming in the Early 1980's, USDA, ERS, Ag. Econ. Report No. 504, September 1983.

Hart, V.B., "Short-Term Borrowing Policies of Farmers," J. Farm Econ. 15:2(April 1933):331.

Hawthorne, H.W., "Some Points Brought Out by Successive Surveys of the Same Farms," J. Farm Econ. 1(June 1919):24-37.

Heady, E.O., Economics of Agricultural Production and Resource Use. Englewood Cliffs, NJ: Prentice-Hall, 1952.

Heady, Earl O., A Primer on Food, Agriculture and Public Policy. New York: Random House, 1967.

Heady, E.O. and R.D. Krenz, Farm Size and Cost Relationships in Relation to Recent Machine Technology, Iowa Agricultural Experiment Station Research Bulletin No. 504, May 1962.

Hibbard, B.H., "A Long Range View of National Agricultural Policy," J. Farm Econ. 16:1(January 1934):13.

Hicks, J.R., Value and Capital, second edition, Oxford: Clarendon Press, 1946.

Hiebert, L. Dean, "Producer Preference for Price Stability," Amer. J. Agr. Econ. 60:1(February 1984):89-90.

Hottel, J. Bruce and Bruce L. Gardner, "The Rate of Return to Investment in Agriculture and Measuring Net Farm Income," Amer. J. Agr. Econ. 65:3(August 1983).

Illinois Farm Business Farm Management Association, Summary of Illinois Farm Business Records, Urbana, IL: University of Illinois Cooperative Extension Service, various years.

Johnson, D. Gale, "International Trade and Agricultural Labor Markets: Farm Policy as a Quasi-Adjustment Policy," Amer. J. Agr. Econ. 64:2(May 1982):354.

Johnson, Glenn L., "An Opportunity Cost View of Fixed Asset Theory and the Overproduction Trap: Comment," _Amer. J. Agr. Econ._ 64:4(November 1982):773-75.

Keynes, J.M., _The General Theory of Employment, Interest, and Money_, New York: Harcourt Brace Jovanovich, 1964.

Lester, Richard A., "Inflation and the Farmer," _J. Farm Econ._ 26:2(April 1934):410.

Limber, R.C., "Conditions Characteristic of Land Booms," _J. Farm Econ._ 20:1(February 1938):233-35.

Lowdon, Frank O., "The Farm Problem," _J. Farm Econ._ 9:1(January 1927):11.

McKenzie, K.J., "Improving Managerial Capabilities of Limited Resource Farmers," _Amer. J. Agr. Econ._ 60:5(December 1978):831-35.

Madden, J. Patrick, _Economies of Size in Farming_. USDA Agr. Econ. Report No. 107, February 1967.

Madden, J. Patrick, "Discussion: What to Do With Those Empty Economic Boxes--Fill or Decorate Them," _Economies of Size Studies_, Center for Agricultural and Rural Development, Ames, Iowa, February 1984.

Madden, J. Patrick and Heather Tischbein, "Toward an Agenda for Small Farm Research," _Amer. J. Agr. Econ._ 61:5(December 1979):940-46.

Martin, William E., "Economies of Size and the 160-Acre Limitation: Fact and Fancy," _Amer. J. Agr. Econ._ 60:5(December 1978):923-28.

Melichar, Emanuel, "A Financial Perspective on Agriculture," _Federal Reserve Bulletin_, (January 1984):1-13.

Melichar, Emanuel, _Agricultural Finance Data Book_. Division of Research and Statistics, Board of Governors of the Federal Reserve System, Washington, DC, July 1984.

Melichar, Emanuel, Personal Communication of Data Series Extended from _Agricultural Finance Data Book_, E. 15(125), Board of Governors of the Federal Reserve System, Washington, DC, September 1984.

Miller, Thomas, "Conceptual Issues in Economies of Size Studies: How Assumptions Drive Away Research," in _Economies of Size Studies_, Center for Agricultural and Rural Development, Ames, Iowa, February 1984.

Miller, Thomas A., Gordon E. Rodewald, and Robert G. McElray, _Economies of Size in U.S. Field Crop Farming_, USDA, ERS, Ag. Econ. Report No. 472, July 1981.

Minsky, Hyman P., _Can "It" Happen Again? Essays on Instability and Finance_. New York: M.E. Sharpe, 1982.

Murray, W.G., "Land Booms and Second Mortgages," _J. Farm Econ._ 20:1(February 1938):230.

Murray, W.G., "Round Table: Can Land Booms Be Avoided?," _J. Farm Econ._ 20:1(February 1938):231.

Murray, William G., "Land Market Regulations," _J. Farm Econ._ 25:1(February 1943):203-18.

Norton, L.J., "The Land Market and Farm Mortgage Debts, 1917-1921," _J. Farm Econ._ 24:1(February 1942):168-77.

Nourse, E.G., "Some Economic Factors in an American Agricultural Policy," _J. Farm Econ._ 7:1(January 1925):1-22.

Olsen, Nils A., "American Agriculture Needs a New Land Policy," _J. Farm Econ._ 11:4(October 1927).

Paarlberg, Donald, "Private Sector Review Loan Guarantes from the Federal Government, Review Team Approach for Most Troubled Loans," Paper presented at the Food and Agricultural Policy Research Institute Workshop on Financial Stress in Agriculture, Kansas City, MO, October 22, 1984.

Patrick, George F. and L.M. Eisgruber, "The Impact of Managerial Ability and Capital Structure on Growth of the Farm Firm," _Amer. J. Agr. Econ._ 50(1968):491-506.

Perkins, Van L., *Crisis in Agriculture: The Agricultural Adjustment Administration and the New Deal 1933 Berkley*. Los Angeles: University of California Press, 1969.

Pershing, Don, Dave Bache, and Freddie Barnard, "What the 1983 Farm Records Show," *Purdue Agricultural Economics Report*, Cooperative Extension Service, Purdue University, West Lafayette, IN, May 1984.

Peterson, George M., "Wealth, Income, and Living," *J. Farm Econ.* 15:3(July 1933):421-51.

Production Credit Association and Farm Credit Service, *AGRIFAX: Comparative Farm Business Analysis*, Fourth Farm Credit District, Louisville, KY, various years.

Raup, Philip, "Economies and Diseconomies of Large Scale Agriculture," *Amer. J. Agr. Econ.* 51(1969):1274-82.

Raup, Philip M., "Some Questions of Value and Scale in American Agriculture," *Amer. J. Agr. Econ.* 60:5(May 1978):303-08.

Rausser, Gordon, "Political Economic Markets: PERTS and PESTU in Food and Agriculture," *Amer. J. Agr. Econ.* 64:5(December 1982):821-33.

Reinsel, Robert D. and Ronald D. Krenz, "Capitalization of Farm Program Benefits into Land Values," ERS-506, United States Department of Agriculture, Economic Research Service, Washington, DC, October 1972.

Rudd, Robert W. and David L. MacFarlane, "The Scale of Operations in Agriculture," *J. Farm Econ.* 24:2(May 1942):420-33.

Rutledge, R.M., "The Relation of the Flow of Population to the Problem of Rural and Urban Economic Inequality," *J. Farm Econ.* 12:3(July 1930):427.

Salontos, Theodore and John D. Hicks, *Agricultural Discontent in the Middle West 1900-1939*. Madison, WI: University of Wisconsin Press, 1961.

Schertz, Lyle P., "Households and Farm Establishments in the 1980s: Implications for Data," *Amer. J. Agr. Econ.* 64:1(February 1982):115-18.

Schultz, T.W., "Theory of the Firm in Farm Management Research," *J. Farm Econ.* 21(1939):57-86.

Schultz, T.W., *The Economic Organization of Agriculture*, McGraw-Hill Book Co., Inc., 1953.

Schultz, Theodore, "Capital Rationing, Uncertainty, and Farm Tenancy Reform," *J. Polit. Econ.* 48(1940):403.

Scott, John T., "Factors Affecting Land Price Decline," *Amer. J. Agr. Econ.* 65:4(November 1983):797.

Shepard, Geoffrey, "Price Discrimination for Ag. Products," *J. Farm Econ.* 20:1(February 1938):792.

Shepard, Lawrence E. and Robert A. Collins, "Why Do Farmers Fail? Farm Bankruptcies 1910-78," *Amer. J. Agr. Econ.* 64:4(November 1982):609-15.

Simpson, Wayne and Marilyn Kapitaniy, "The Off-Farm Work Behavior of Farm Operators," *Amer. J. Agr. Econ.* 65:4(November 1983):801-05.

Sporleder, Thomas L., "Emerging Information Technologies and Agricultural Structure," *Amer. J. Agr. Econ.* 65:2(May 1983):388-94.

Stanton, B.F., "Perspective on Farm Size," *Amer. J. Agr. Econ.* 60:5(December 1978):727-37.

Stigler, G.H., "The Economies of Scale," *J. Law Econ.* 1(1958):54-71.

Stigler, George J., "Social Welfare and Differentiated Prices," *J. Farm Econ.* 20:1(1938):573-86.

Summary of Illinois Farm Business Records, University of Illinois at Urbana-Champaign, Cooperative Extension Service Circular No. 1229, West Lafayette, IN (various annual reports).

Suter, Robert C., "A Summary of the Indiana Farm Accounts 1976-1982: All Farms," Mimeo No. 0053P-0015P, Department of Agricultural Economics, Purdue University, December 26, 1953.

Taylor, H.C., Introduction to the Study of Agricultural Economics. New York: Macmillan Co., 1905.

Taylor, H.C., "The Adjustments of the Farm Business to Declining Price Levels," J. Farm Econ. 3:1(January 1921):

Thompson, E.H., "Can Land Booms Be Avoided?," J. Farm Econ. 20:1 (February 1938):235-36.

Thurow, Lester C., Generating Inequality: Mechanisms of Distribution in the U.S. Economy. New York: Basic Books, 1975.

Timmons, John F., "Farm Ownership in the United States: An Appraisal of the Present Situation and Emerging Problems," J. Farm Econ. 34:1(January 1948):78-100.

Tolly, H.R., "Objectives in National Agricultural Policy," J. Farm Econ. 20:1(February 1938):

Tosterud, Robert J., "Commodity Programs: Discussion," Amer. J. Agr. Econ. 65:5(December 1983):932-34.

Tweeten, L., "Farm Commodity Prices and Income in the Farming Sector," Consensus and Conflict in U.S. Agriculture, ed. B. Gardner and J. Richardson. College Station, TX: Texas A&M University Press, 1979.

Tweeten, Luther, Foundations of Farm Policy, Lincoln, NB: University of Nebraska Press, 1970.

Tweeten, Luther, "Economic Instability in Agriculture: The Contribution of Prices, Government Programs, and Exports," Amer. J. Agr. Econ. 65:5(December 1983):922-31.

Tweeten, Luther G., "Macroeconomics in Crisis: Agriculture in an Underachieving Economy," Amer. J. Agr. Econ. 62:5(December 1980):853-65.

Tweeten, Luther G., "Typology and Policy for Small Farms," *Southern J. Agr. Econ.*, December 1980.

Tweeten, Luther, Bruce Gardner, Emerson M. Babb, William D. Heffernan, Marshall A. Martin, James Richardson, Bernard Stanton, and Fred C. White, *The Emerging Economics of Agriculture: Review and Policy Options*. Ames, Iowa: Council for Agricultural Science and Technology, Report No. 98, 1983.

United States Department of Agriculture, *Agriculture Statistics*, United States Government Printing Office, Washington, DC, various issues.

Upchurch, M.L., "Implications of Economies of Scale to National Agricultural Adjustments," *J. Farm Econ.* 43(1961):1239-46.

USDA, *A Time to Choose: Summary Report on the Structure of Agriculture*, Washington, DC, January 1981.

USDA/ERS, "Financial Conditions of Farmers and Farm Lenders," Roosevelt Center for American Studies, Washington, DC, November 27, 1984.

Warren, G.F., "Which Does Agriculture Need--Readjustment or Legislation?," *J. Farm Econ.* (January 1928):1.

Warren, G.F. and F.A. Pearson, *Gold and Prices*. New York: John Wiley and Sons, Inc., 1935.

Warren, Norman J., "Developments with Respect to Short Term and Emergency Credit," *J. Farm Econ.* 15:2(April 1933):37.

Waugh, Fredrick V., "Market Prorates Social Welfare," *J. Farm Econ.* 20(1938):403-16.

Wehrwein, George S., "Problem of Inheritance in American Land Tenure," *J. Farm Econ.* 9(1927):163.

T - #0178 - 071024 - C0 - 224/148/12 - PB - 9780367163747 - Gloss Lamination